Science, Religion, and the Human Future

Conflict, Collusion, and Consequences

AMANDA REES
University of York

FRANZISKA E. KOHLT
University of Leeds

TOM McLEISH
University of York

CHARLOTTE SLEIGH
University College London

DAVID WILKINSON
St Johns College, Durham University

OXFORD
UNIVERSITY PRESS

OXFORD
UNIVERSITY PRESS

Great Clarendon Street, Oxford, OX2 6DP,
United Kingdom

Oxford University Press is a department of the University of Oxford.
It furthers the University's objective of excellence in research, scholarship,
and education by publishing worldwide. Oxford is a registered trade mark of
Oxford University Press in the UK and in certain other countries.

© Amanda Rees, Franziska E. Kohlt, Tom McLeish,
Charlotte Sleigh, David Wilkinson 2025

The moral rights of the authors have been asserted.

All rights reserved. No part of this publication may be reproduced, stored in a retrieval system,
transmitted, used for text and data mining, or used for training artificial intelligence, in any form or
by any means, without the prior permission in writing of Oxford University Press, or as expressly
permitted by law, by licence or under terms agreed with the appropriate reprographics rights
organization. Enquiries concerning reproduction outside the scope of the above should be sent
to the Rights Department, Oxford University Press, at the address above.

You must not circulate this work in any other form
and you must impose this same condition on any acquirer.

Links to third party websites are provided by Oxford in good faith and
for information only. Oxford disclaims any responsibility for the materials
contained in any third party website referenced in this work.

Published in the United States of America by Oxford University Press
198 Madison Avenue, New York, NY 10016, United States of America

British Library Cataloguing in Publication Data
Data available

Library of Congress Control Number: 2025946070

ISBN 9780198889007

DOI: 10.1093/9780191995316.001.0001

Printed and bound by
CPI Group (UK) Ltd., Croydon, CR0 4YY

MIX
Paper | Supporting
responsible forestry
FSC® C013604

The manufacturer's authorized representative in the EU for product safety is
Oxford University Press España S.A. of Parque Empresarial San Fernando de Henares,
Avenida de Castilla, 2 – 28830 Madrid (www.oup.es/en or product.safety@oup.com).
OUP España S.A. also acts as importer into Spain of products made by the manufacturer.

For our friend, colleague, and mentor Professor Tom McLeish, who embodied an exciting and joyful engagement of science and Christian faith.

Introduction

Myth and more

This is how the myth goes:

Once upon a time, a long, long time ago, humanity was afraid.

Humans' fears grew out of ignorance. They did not know how to combat drought or disease. And so, they created superstitions and rituals to placate the gods whom they thought brought this suffering upon them, wasting their capacity to think clearly and build better worlds. Those heroic few who were able to break free from bonds of superstition found themselves under attack: sunlight can blind and terrify those accustomed only to see the shadows on the wall of the cave...

Then, at one particular time, and in one particular place, a revolution began to take shape. A few men in sixteenth-century Europe rejected irrational superstition and focused instead on observable proof. Rather than taking on trust the wisdom of the ancients, or accepting without question the word of God, people like Nicolaus Copernicus, Francis Bacon, Giordano Bruno, Rene Descartes, Galileo Galilei, and Isaac Newton began to test and understand the world around them. From their endeavours grew the 'Scientific Revolution', a feat of human ingenuity achieved in the teeth of opposition from the church. Science gave humanity the tools to manipulate and control the natural world around itself, ultimately enabling travel to the heavens themselves.

The first battles between science and religion had been won; yet even in the twenty-first century, the war is not yet over. Rationalism must remain vigilant if it is to ward off the encroachments of superstition . . .

This myth resonates powerfully through modern Western culture. It describes a conflict between science and religion that is not just deeply rooted in human history but even inevitable and universal. Science is depicted as a matter of objective reason, religion as one of subjective belief.

Science, Religion, and the Human Future. Amanda Rees et al., Oxford University Press. © Amanda Rees, Franziska E. Kohlt, Tom McLeish, Charlotte Sleigh, David Wilkinson (2025).
DOI: 10.1093/9780191995316.003.0001

These basic definitions are fleshed out in a series of additional clichés. Science, so it is said, gives access to observable and accurate reflections of the world around us. By contrast, religion is supposed to be based in blind faith or trust in received authority. The 'scientific method' is often cited as an anchor for science's impartiality and accuracy. People who talk about this usually mean the testability of hypotheses, or the replicability of scientific experiments. There is, so the complementary cliché goes, no testing of religion. Science is supposed to have no ethical or moral implications in and of itself; only its *application* can sometimes be problematic. On the other side of this coin, some people are content to assert that religion is *always* harmful or imposed by its leaders out of selfish intent. Finally, science is often (though not always) portrayed as an activity that does not require emotion or imagination, whereas these qualities are supposedly the only things that inform religious experience. Faith, according to these definitions, must always be opposed to reason, just as ignorance is opposed to knowledge. For rationalists, science represents a fundamental obstacle to the success of religion; for fundamentalists, it is the other way around. Ironically, and dangerously, the uncritical use of this myth to shape public discourse has often resulted in the undermining of the very science it seeks to promote.

As many historians have attested, the facts of history do not bear out the story sketched out above. Following their lead, this book will show that the 'conflict' myth is based in an imperfect understanding, and sometimes a conscious misinterpretation, of history. It is a myth that was deliberately created in the late nineteenth century to meet the particular political needs of a particular historical moment. It is a myth that maintains cultural traction in the present day because it continues to serve the interests of particular groups and institutions. Its use in the present is moreover doing great harm to public discussions of knowledge, expertise, morality, and responsibility.

This book will argue that it is vital to our collective future that we treat science and religion not as opposed or even as separate domains, but as shared human endeavours. Both are activities that are carried out by individuals and communities, in interaction with each other, and with other aspects of the world, both physical and metaphysical. The two activities do not exist independently of human societies: instead, they are conducted through behaviours and motivations that represent critical components of our common human heritage. They are grounded in a questioning, engaged, and hands-on approach to the world around us, whether that is reflected in hymns and equations or embodied in cave painting and Large

Hadron Colliders. Recognising and acting on this fact is essential: this book represents our efforts so to do.

We are not, of course, the first to question the conflict myth, and we won't be the last. It has been argued that continuing to discuss the 'conflict', even for the purposes of defusing it, perpetuates its power.[1] Meanwhile, an excellent recent collection of articles on the topic edited by the zoologist Jeff Hardin and his colleagues describes this myth as 'the idea that wouldn't die'.[2] The conceptual, empirical, and moral relationships between science and religion have been the subject of sustained critical scholarship ever since its first appearance in the 1870s.[3]

Where this book differs from its predecessors is, first, in our diagnosis of a hidden and sometimes unhealthy *collusion* between science and religion in modern times and, second, in our exploration of the polarising *consequences* that a conflict framing of the science/religion relationship has had for Anglo-American culture. Its chapters explore a range of key public debates and seek to identify new tools and strategies for understanding the relationship between these two activities. The rest of this introduction, therefore, will position this book within key themes in the existing work on the intellectual and political development of the relationship between science and religion, as well as outlining the structure and content of the remaining chapters. At the outset, though, it's important to begin by at least *trying* to define terms.

Science and religion in history

What do we mean by science and religion? To some, the answer to this question will seem obvious. Religion implies a system of commonly held beliefs and practices, usually or often involving the worship of an immaterial divine power. Science, on the other hand, refers to an organised body of knowledge and skills, based on the use of observation and experiment to systematically study the material world. In the Anglo-American world, the term would not normally include the arts, humanities, or social sciences.[4] Most readers will probably want to tweak or to challenge either (or both) of those definitions but broadly speaking would recognise them as accurate descriptions of how the two are commonly understood. Neither of them, however, accurately describes the activities and beliefs of cultures outside the modern West.

The meanings currently attached to science and religion are not only culturally specific but are also relatively modern. During the twelfth and thirteenth centuries, scholars in Islam (Al-Ghazâlî, Ibn Rushd), Judaism

(Maimonides), and Christianity (Aquinas) all wrestled with the implications of Aristotle's natural philosophy, often considered as an early kind of science. They found points of complementarity and points of tension with their respective religions. But none of them condemned his overall philosophy as a heretical way of thinking: they did not consider it as a complete or exclusive system of knowledge in the way that science is now commonly regarded.

Historian Peter Harrison notes that the roots of the words 'science' and 'religion'—*scientia* and *religio*—refer not to external activities, behaviours, or belief in sets of propositions, but to states of internal virtue.[5] For Aristotle and other ancient Greek philosophers, the acquisition of knowledge was essential to pursuing excellence and living one's best life. Importantly, however, this was not knowledge for its own sake. Instead, what mattered was the process of acquiring it. It was a means of exercising one's mind to prepare it for the subtle contemplation of complexity. Logical demonstration, in this context, was an exercise that aimed at the transformation and the improvement of the self. For medieval scholars, the study of God's word and works, whether written in scripture or evident in the complexity of the natural world, required sensitivity to allegory and a capacity for hermeneutic deliberation (interpretation). Neither literal interpretation of the Bible nor attention to purely surface appearance would suffice if a true understanding of the object of study was to be apprehended.

In ancient and medieval cases, the activities and motivations which modern writers have retrospectively identified as scientific or religious were understood by their practitioners in a way that differed profoundly from the modern meaning. *Scientia* and *religio* represented closely connected attributes of the individual which aimed at preparing them to live a better life. Historians such as David Lindberg and Ronald Numbers have been in the forefront of efforts to refute the notion that the medieval Christian church suppressed the study of the physical world. They have repeatedly debunked myths such as the claim that the church banned human dissection or taught that the Earth was flat.[6] As Chapter 1 of this book will show, theology and faith did not only motivate natural philosophers to examine and apprehend the external world; it also gave them the key tools to analyse it.

Even after the Reformation had apparently stripped the world of its enchantment, by encouraging people to take a more literal approach to the interpretation of both scripture and the natural world, theology and natural philosophy continued to inspire one another. Galileo may have used mathematics rather than allegory as the principal tool through which he understood the world, but this did not detract from his belief that the natural laws identified through observation and described by mathematics had been imposed on the world by God, who remained the first and efficient cause of everything. Regularities and connections in the physical world were the direct manifestation of God's activities: observing them was to observe part of God's action in the world. This perspective was at the heart also of Newtonian philosophy; Newton was a keen if unorthodox theologian.[7]

Ultimately, the development of the experimental method was meant to compensate for the fact that, unlike God, humans could and would make mistakes. Telescopes and microscopes improved on limited eyesight; methods of recording and calculation improved on fallen minds. Early natural philosophers increasingly came to feel that it was through the disciplining and control of the natural world, rather than through mastering one's internal self, that fallen humanity could recover its relationship with God. In this way, early modern natural philosophers could be seen as secular theologians: for practitioners of physico-theology, reason now could support faith. Theology could be regarded (as it was until mid-nineteenth century British university curricula) as one of the inductive sciences, that is, reasoning from evidence to truth.

By the time of the enlightenment, representatives of an emergent 'modern' science sought to translate its ancient virtues into a collective public good. In particular, scientific ideas about progress, improvement, and development were critical to the rapid development of industrial capitalism and the ruthless expansion of imperialism in Europe and North America. Together, these cultural innovations provided a narrative by which the West could position itself at the apex of civilisation, justifying imperial projects that brought its benefits to other parts of the world. The result was a triumphalist story of Western civilisation, based on the construction of a universal, value-free, and rational philosophy. To a considerable degree, this rhetoric equated the idea of civilisation with Christianity. While missionaries continued to proselytise amongst the 'heathen', the broader 'civilising mission' of Europe focused upon the benefits brought by railways, schools, hospitals, and other aspects of scientific, technological, or medical development. Postcolonial scholars from Frantz Fanon to Ruha Benjamin have dissected and demonstrated the manifold errors and deliberate omissions in this position.[8] The railways of India, for example, were positioned to facilitate exports for British profit rather than to enhance the domestic economy. The sciences of anthropology were used to divide and suppress Indigenous populations. The necessity of a civilised education was used as a rationale to separate children from their parents. Imperial science—the 'civilising mission'—resulted in manifest injustice and inhumanity.

Postcolonial and Indigenous scholars have critiqued the idea that science is inherently Western. This description actively excludes the essential intellectual and material contributions made to modern science by cultures and communities other than 'Western' ones. These contributions, in both past and present, have been frequently omitted or glossed over in standard histories of the period.[9] The Islamic Golden Age, for example, represented one of the great, sustained periods of human creativity in history (see Chapter 1).[10] Stretching in time from the eighth to the thirteenth century CE, it played an essential role in laying the foundations for the 'European' scientific revolution. Yet despite its familiarity to historians, this part of the story of science rarely appears in public debates.

By the later nineteenth century, the public benefits of science were allied with a defined set of institutions and practices that governed its pursuit. Degrees, laboratories, disciplinary societies, and professional publications all helped to define what counted as science. Alongside these institutional elements, standards of conduct were developed. These have been described as a kind of aspiration to record the phenomena of nature as though one were a machine, a kind of 'mechanical objectivity'.[11] Disinterestedness was to be cultivated. Emotion was to be shunned. Upon this basis, men of science could be trusted to bring their findings and machines to society. The word 'scientist' was invented to account for the similarities of attitude and practice that were supposed to unite the chemist, the geologist, and the botanist. First coined in the 1830s, it found common currency around the time that the conflict thesis was born.

The birth of a myth

Historians of science agree that the notion of a conflict between science and religion appeared as a public concept in the late nineteenth century.[12] It emerged at the point when British scientists were beginning to develop a self-conscious collective identity and to seek wider cultural recognition of their authority. They wanted to have control over what children were taught in school; control over what values guided society and its governance. In a country such as Britain, where bishops still held sway in the House of Lords, Christianity had to step aside to make space for their ambitions. In a rhetorical *coup de grâce*, these scientists were able to argue that their 'objective' science was of a superior philosophical status to the purely 'subjective' morality of their rivals. From this time onwards, the 'conflict' narrative and its retrospective mapping onto human history increasingly characterised the way that individuals thought about the relationship between faith and the natural world.

The claim of scientists to lead their nation echoed the manifestos of earlier nineteenth-century philosophers, ranging from Auguste Comte and Emile Durkheim to Karl Marx and Fredrich Engels, who had adopted a progressive approach to their analysis of human history. They had described the stages through which societies evolved in favour of rationality, technology, and industrial civilisation.[13] Re-read through the lens of Darwinian evolution, the atrophy or rejection of religion as a part of this process gained increased prominence. Religion came to be synonymous with primitive stages of social development, immaturity, and the past: science represented mature adulthood, self-control, and the future. Narratives of scientific progress which had emerged in a Christian context now became a part of the 'conflict' narrative. Social development, read through the lens of Darwinian evolution, became and remains the banner of anti-clerical science.

In 1874, John Tyndall gave a speech to British scientists which crystallised the conflict narrative, placing present-day science as the culmination of a straight line of progress from pagan Greek philosophy through the Renaissance, triumphing despite the best efforts of the medieval Church to derail and suppress it.[14] Meanwhile, Darwin's close friend and supporter T. H. Huxley formed his so-called 'X Club' in order to discuss how science could be separated from theology. It comprised a group of highly influential scientists who actively worked to remodel the Royal Society along this separatist line, even though even Huxley's own views, as expressed in private correspondence, were more complex.[15] As science became the domain of professional practitioners, clergy were excluded from its institutions, becoming (alongside other amateurs and all women) a new kind of laity, unwelcome to contribute to scientific debates.

Instead of being guided by values channelled through the words of bishops and priests, the cognitive authority of science now entitled scientists to speak with authority. They could provide audiences with the answers to the questions that troubled them. Astronomy and evolution were, and perhaps remain, amongst the most common topics in popular science: secular stories that tell us about where we came from and who we are. For some twentieth-century biologists—Julian Huxley, J. B. S. Haldane—genetic sciences in particular would enable humanity to take control of their own destiny. Through advances in biology and cybernetics, the human mind would be enabled to transcend the human body (Chapter 7).

From this time onwards, the 'conflict' narrative and its retrospective mapping onto human history increasingly characterised the way that individuals thought about the relationship between faith and the natural world.

Non-overlapping magisteria

Many writers and commentators have tried to resolve the conflict by arguing that science and religion are different either in their topics of interest, or in their methods, or both. They separate them into the natural world versus the supernatural, for example, or empirical observation versus internal contemplation. One of the most popular formulations of this response suggests that science tells us 'how' while religion tells us 'why'. A closely related variant on this theme holds that science is concerned with facts, while religion is concerned with values.

The evolutionary biologist and historian of science Stephen Jay Gould once reflected on a discussion he had had with a group of Jesuit priests bemused by the growth in popularity of 'scientific creationism' in the US. One concerned member of the group protested that:

> I have always been taught that no doctrinal conflict exists between evolution and Catholic faith, and the evidence for evolution seems both entirely satisfactory and utterly overwhelming. Have I missed something?[16]

Gould, aware of the irony of an agnostic Jew advising a group of Catholic clergy, reassured his audience that creationism was merely a local American phenomenon, produced by and for fundamentalist Protestant sects committed to a literal reading of the Bible.

From this experience, Gould went on to write an essay that articulated one of the best-known versions of the distinction between science and faith. He cited Pope John Paul II's 1996 declaration that evolution was not a hypothesis but a fact as the basis for developing his own position.[17] No 'war' could exist, Gould suggested, because the two forces were not trying to occupy the same territory. Science, he argued, dealt with the material world—how it was made and how it worked. In contrast, religion dealt with morality and value. Natural selection could produce the human body, but it was the divine that infused the human soul. Science and religion therefore each embraced very different ways of approaching their worlds; they dealt with distinctive topics and adopted distinctive methodologies. They gave rise to very different worldviews, each governed by its own rules that delineated *what* was known and even what *could* be known. They were, in Gould's terms 'non-overlapping magisteria'.

Gould's proposal met with some critical responses. His colleague at Harvard, the astronomer and historian Owen Gingerich, pointed out that there were quite a few cases where the interests and concerns of science and religion overlapped directly and considerably. The obvious examples here are the cases of the Copernican and Darwinian hypotheses: the reception given to both ideas by philosophers, scientists, theologians, and priests are, of course, well-known stories within the broader conflict narrative. But Gingerich's third example, the theory of how chemical nuclei are made within stars, was less familiar. The theory's author, astronomer Fred Hoyle, pondered how it was that carbon nuclei come out exactly as necessary to make possible the existence of life. (All living organisms are built on carbon molecules.) Gingerich quotes Hoyle as asking:

> Would you not say to yourself, 'Some super-calculating intellect must have designed the properties of the carbon atom, otherwise the chance of my finding such an atom through the blind forces of nature would be utterly minuscule?' A common sense interpretation of the facts suggests that a super-intellect has monkeyed with physics, as well as with chemistry and biology, and that there are no blind forces worth speaking about in nature.[18]

It's a short step here from atomic properties to the existence of a deeper story to the universe. A detailed discussion of the anthropic principle is beyond the scope of this introduction. But the existence of the debates themselves demonstrate that one can only successfully separate the domains of science and religion by introducing artificial boundaries that are in fact highly permeable.[19]

Gould's division of labour and powers, while intending to reduce potential for conflict, relied upon the modern and culturally specific assumptions that underpinned the making of the myth. It reinforced the idea that science and religion exist as predetermined categories, rather than as words that describe activities carried out by blurred and mixed human communities. As Gould himself noted, the two domains could sometimes be too close for comfort. To what extent is it possible to leave morality outside the laboratory? It would require a thick set of blinkers for scientists not to guess at some of the likely applications of their research.

The philosopher Max Horkheimer wrote extensively about the dangerous consequences that would follow from adopting a scientific position of asking '*how* can x be done?' (instrumental reason), rather than '*should* x be done?' Horkheimer linked the growth of instrumental reason in science to the rise of fascism: rationalism combined with technocracy creates a hyperfocus on solutions rather than moral consequences.[20] Similar anxieties can be identified in relation to the rise in later twentieth-century concern for laboratory and livestock animal welfare, the genetic modification of organisms, experimentation with human embryos, and the racialised impact of technological innovation. It is not possible to detach morality from practice in these cases—and not all scientists would try to do so.

Even if it *were* possible to leave morality outside the lab, what impact would that have on the work done by those within it? As we shall see in Chapter 11, feminist philosophers of science have been particularly instrumental in highlighting how the inevitable gap between evidence and hypothesis confirmation is bridged by assumptions and beliefs. Gould's attempt to sidestep the 'conflict' question by suggesting that 'science tells you *how* to do things, religion decides if they are *good* things to do' leaves intact the notion that objectivity is a morally neutral stance. In fact, as the chapter on climate change explores, the making of the scientific subject (that is, the person who possesses objectivity) involves the active erasure of empathetic identification with the people or things that one is studying. Such a God's-eye-view definition of objectivity is poor Christian theology which, by contrast, premises a God who compassionately identifies with creation.

What this book is about

Science, as this introduction has stressed, does not have an existence independent from its social context, but is an activity undertaken by human beings in interaction with each other and with the world around them. As such, it is constantly subject to disagreement, controversy, and revision. Indeed, pretending that science is not like this can backfire on scientists who have staked their reputation on its inherent reliability, only to find that they have made a mistake.[21] Conspiracy theories thrive in cracks such as these. The following chapters re-situate stories of science—past, present, and future—in their social context, showing how fact and value, science and faith, are interwoven at every level.

Deciding how we want science to serve us in the future demands a multi-perspectival view that includes values and faith. This is why we have written this book. As we have noted, many of our colleagues have already covered in detail the errors of fact and interpretation that make up the 'conflict' myth and have done an outstanding job of exploring the evolving relationship between 'science' and 'religion' throughout human history. Some of these investigations can be found under the further readings listed at the end of each chapter. What we want to do is to show how the conflict myth frames a rhetorical paradigm that impacts directly and detrimentally on present-day debates. At the same time, we demonstrate how Christian theology often underpins the scientific narrative and choice of metaphor. In five case studies, we identify and examine how and when this happens, and what the consequences are. We will also suggest strategies whereby public debates about knowledge, morality, and expertise can be carried out more effectively, and how we can identify and create alternatives to these artificial, distorted, and polarising narratives.

For better and for worse, this book focuses on Christianity. The book has emerged from a project centred in this faith tradition (Equipping Christian Leadership in an Age of Science, ECLAS) and this is where our knowledge and expertise, such as they are, are located. To some extent, this equips us well. Christianity formed the major cultural and religious backdrop for many of the scientific episodes that are discussed within these pages. Chapter 1 describes medieval science that is deeply rooted in Hellenic and Islamic philosophy, but by the time our history reaches the enlightenment, it mostly concerns the intersection of science and Christianity. This, more than any other religion, shaped imperialist and colonial activity in relation to 'progress' and 'civilisation'. Christianity also, therefore, formed the cultural backdrop for the creation of the conflict myth with which this chapter opened. Modern Christianity and modern science were the two garments cut from the cloth of the seventeenth century, by inhabitants of nineteenth-century Britain and the US. The authors' expertise in Christian theology also enables them to probe some of the specific metaphors and narratives that constitute the scientific case studies in Part II. We find many echoes of Christian theology within them, or perhaps a more apt metaphor would be 'undigested remains of poor Christian theology'. Chapters 10 and 11 eventually touch upon growing scholarship in relation to science and other faiths, but inadequately so. We look forward to becoming better educated through the interfaith work in science that is currently underway.

We have divided the book into three sections. The first concentrates on some important aspects of the history of the relationship between science and religion. Chapter 1 focuses upon the ancient and medieval period to demonstrate the intimate connections between theology, philosophy, and the investigation of the natural world amongst Islamic Golden Age and medieval scholars. Chapter 2 considers the evolving relationship between science, Christianity, capitalism, and imperialism in the eighteenth and nineteenth centuries, with an emphasis on the confluence of science and religion in creating a joint narrative about progress and civilisation. This common ground, generally overlooked in accounts of the conflict, is of significance in Part II of the book. Chapter 3 gives an account of the broader cultural context in which the 'conflict thesis' was generated. It finds a parallel between the nineteenth-century polemicists and Thomas Sprat, seventeenth-century historian of the Royal Society, in their desire to attack Catholicism in particular. It describes how an initially sectarian focus gradually grew to encompass Christianity and then religion as a whole in Anglophone cultures. Most recently, the myth resurfaced during the 'Science Wars' and the increasing polarisation of public cultures in the late twentieth and early twenty-first centuries.

The second part of this volume focuses on case studies from recent and contemporary history, showing how narratives from the conflict thesis underpin them. What emerges from most, if not all, of these examples is that theology, so far from conflicting with science, has informed much of its tacit and explicit narrative framing. Bad theology, of the sort that co-created the myth of civilisation and progress, has been baked into the public discourses surrounding science and technology.

Our first case study, in Chapter 4, examines religion and the human future in space. It explores how theology, religion, and faith have supported humanity in reaching for the stars: providing cosmological insight, a motive for technological innovation, and emotional and spiritual succour for those venturing beyond the atmosphere. It is the first of several chapters to define faith and science in close proximity as frameworks to imagine and pursue a preferred vision of the future. Chapter 5 focuses on the debates surrounding the genetic modification of organisms and reveals the historical and geographic specificity of the 'conflict thesis'. Concerns about 'playing God' and 'editing the book of nature' may dominate public discussion in the West, but very different stories are told elsewhere, prompted by perceptions of inequity. These critiques refuse the fact/value division of science and faith, revealing that if there is guilt in regard to 'playing God', the offence occurs upstream of GMO (genetically modified organism) release, in the posture of scientific objectivity. Again, we see the envisioning of possible futures; in this instance, critics have highlighted the limitations of imagination amongst Western scientists, who do not realise that different communities will imagine different kinds of preferred futures.

Chapter 6, on climate change, pinpoints the way in which early debates about the relationship between science and policy hinged on the question of 'belief'. This thin theology is argued to be a legacy of the conflict thesis, which demands a binary choice between belief in either science or religion. Faith, conceived as the practical orientation of imagination, is argued to be a much better description of the kind of knowledge that may generate an improved climate future. As such, it can apply to science or religion. Artificial intelligence is the topic of Chapter 7. It reveals how theological and faith-based metaphors—technological versions of the saved and the damned—permeate the AI debate, echoing narratives from the joint civilising mission described in Chapter 2. It describes the ways in which the projected rise (and fear) of autonomous self-willed machines can be understood in the context of the relationship between morality and rationality. Turning to other recent events, Chapter 8 addresses the recent Covid-19 pandemic. It is about the dangers of war narratives in general and examines the interplay of science and faith at several levels of communication during the outbreak. Religious leaders were anxious to demonstrate their credentials in supporting rational solutions, solutions which themselves echoed aspects of the civilising mission described in Chapter 2.

In the third and final part of this volume, we focus our attention on the present and the future. Chapter 9 draws out some of the most significant parallels between aspects of scientific and religious cultures as they are represented and perceived in popular culture. It shows how scientists are consistently presented as fundamentally different from 'normal' people. We critically examine some of the criteria for scientific 'canonisation', yielding tools to rethink the presentation of science as omniscient and omnipotent. Chapter 10 examines a late flash in the science–religion conflict, the new atheist movement of the early twenty-first century. We set it in the context of social trends which do not support a thesis of growing global secularisation, and in the context of a growing phenomenon of scientist-theologians. Chapter 11 broadens the context still further, taking in the work of scholars in Science and Technology Studies who have undermined the myth of scientific objectivity from other angles. Their work complements the efforts of Indigenous activists who advocate ways of knowing that entangle subject and object, fact and value. These examples yield prompts for alternative narratives, metaphors, and strategies for talking about science in this precarious era. Our futures, after all, depend upon humanity's capacity to find better ways of managing knowledge, expertise, and values. A concluding chapter describes the attempts of ECLAS to put some of these into practice.

Our argument throughout is that neither science nor faith is, or should be, the sole province of priests or scholars. Both are human activities, and the questions that motivate them are profoundly human. They are constituted from actions and beliefs in which we all participate, knowingly or otherwise. There are theologies in both, whether we like it or not. Theologies cannot be extracted, but they can be improved. To build better human futures, we urgently need to reclaim these theologies.

To do this, we turn, first, to history.

Further reading

Thomas Dixon and Adam Shapiro, *Science and Religion: A Very Short Introduction*, 2nd edn (Oxford: Oxford University Press, 2022)

Owen Gingerich, *God's Planet* (Cambridge, MA: Harvard University Press, 2014)

Jeff Hardin, Ronald L. Numbers, and Ronald A. Blazey, *The Warfare Between Science and Religion (The Idea That Wouldn't Die)* (Baltimore: Johns Hopkins University Press, 2018)

Peter Harrison, *The Territories of Science and Religion* (Chicago: Chicago University Press, 2015)

David Hutchings and James C. Ungureanu, *On Popes and Unicorns: Science, Christianity, and How the Conflict Thesis Fooled the World* (Oxford: Oxford University Press, 2022)

Ronald Numbers, ed., *Galileo Goes to Jail and Other Myths about Science and Religion* (Cambridge MA: Harvard University Press, 2009)

Notes

1. Rebecca Catto, James Riley, Fern Elsdon-Baker, Stephen H. Jones, and Carola Leicht, 'Science, Religion, and Nonreligion: Engaging Subdisciplines to Move Further Beyond Mythbusting', *Acta Sociologica* 66.1 (2023): 96–110.
2. Jeff Hardin, Ronald L. Numbers, and Ronald A. Blazey, *The Warfare Between Science and Religion (The Idea That Wouldn't Die)* (Baltimore: Johns Hopkins University Press, 2018).
3. Alistair McGrath, *Science and Religion*, 3rd ed. (Oxford: Blackwell, 2020); James C. Ungureanu, *Science, Religion and the Protestant Tradition: Retracing the Origins of Conflict* (Pittsburgh: University of Pittsburgh Press, 2019); John Hedley Brooke, *Science and Religion: Some Historical Perspectives* (Cambridge: Cambridge University Press, 2014);

David Hutchings and James C. Ungureanu, *On Popes and Unicorns: Science, Christianity, and How the Conflict Thesis Fooled the World* (Oxford: Oxford University Press, 2022).

4. In other languages, such as German, the term 'science' can include these other fields of human knowledge.

5. Peter Harrison, *The Territories of Science and Religion* (Chicago: Chicago University Press, 2015).

6. David C. Lindberg and Ronald L. Numbers, eds, *When Science and Christianity Meet* (Chicago: Chicago University Press, 2003). Sacrobosco's *De Sphaera* (The Sphere) was a key resource for medieval scholarship.

7. Rob Iliffe, *Priest of Nature: The Religious Worlds of Isaac Newton* (Oxford: Oxford University Press, 2016).

8. Frantz Fanon, *A Dying Colonialism* (New York: Grove Press, 1965); Ruha Benjamin, *Race after Technology: Abolitionist Tools for the New Jim Code* (Oxford: Wiley and Sons, 2019).

9. Benjamin Elman, *A Cultural History of Modern Science in China* (Cambridge: Harvard University Press, 2006); Sheldon Pollock, *The Language of the Gods in the World of Men: Sanskrit, Culture, and Power in Premodern India* (Berkeley: University of California Press, 2006); Karen Thornber 'Humanistic Environmental Studies and Global Indigeneities', *Humanities* 5.3 (2016): 52, https://doi.org/10.3390/h5030052; James Poskett, *Horizons: A Global History of Science* (London: Penguin, 2022); Clapperton Mavhunga, ed., *What Do Science, Technology, and Innovation Mean from Africa?* (Cambridge, MA: MIT Press, 2017).

10. A. I. Sabra, 'Situating Arabic Science' in *The Scientific Enterprise in Antiquity and the Middle Ages*, edited by Michael H. Shank (Chicago: Chicago University Press, 2000).

11. Lorraine Daston and Peter Galison, *Objectivity* (New York: Zone Books, 2007).

12. David B. Wilson, 'The Historiography of Science and Religion' in *Science and Religion: A Historical Introduction*, edited by Gary B. Ferngren (Baltimore: Johns Hopkins University Press, 2002).

13. Raewyn Connell, *Southern Theory: Social Science and the Global Dynamics of Knowledge* (Cambridge: Polity Press, 2007).

14. Roland Jackson, *The Ascent of John Tyndall: Victorian Scientist, Mountaineer and Public Intellectual* (Oxford: Oxford University Press, 2018).

15. Ruth Barton, *The X Club: Power and Authority in Victorian Science* (Chicago: Chicago University Press, 2018).

16. Stephen J. Gould, 'Non-Overlapping Magisteria', *Natural History* 106 (1997): 16–22 and 60–2, p. 16.

17. Pope John Paul II, 'Truth Cannot Contradict Truth', 22 October 1996. Contrast with Pope Pius XII's declared position in Humani Generis, 12 August 1950.

18. Owen Gingerich, *God's Planet* (Cambridge MA: Harvard University Press, 2014), p. 126.

19. John D. Barrow and Frank J. Tipler, *The Anthropic Cosmological Principle* (Oxford: Oxford University Press, 1988).

20. Max Horkheimer, *The Eclipse of Reason* (Oxford: Oxford University Press, 1947).

21. Niels Mede and Mike Schäfer, 'Science-Related Populism: Conceptualizing Populist Demands toward Science', *Public Understanding of Science* 29.5 (2020): 473–91, https://journals.sagepub.com/doi/full/10.1177/0963662520924259.

PART I

PAST

1

An ordered universe

A single cloth

We saw in the introduction that the categories of 'science' and 'religion' should be applied to the pre-modern period with great caution. These terms are understood to apply to bodies of knowledge, concepts, methodologies, and practices that are communally held by human communities, while the *scientia* and *religio* of earlier centuries referred to internal virtues possessed by an individual. This chapter explores how knowledge of the world was created and understood by medieval people living in Western Europe, demonstrating that theology and natural philosophy were pursuits cut from a single cloth: followed in the same places, at the same times, and often by the same people. It will show how the institutions of the Catholic Church, from the Papacy to local monastic schools, created spaces within which knowledge of the natural world, and the methods through which this knowledge could be acquired, were pursued and contested. Before the later nineteenth century, 'religion' was neither at war nor peace with 'science': instead, theologians, natural philosophers (and sometimes bureaucrats) explored the boundaries of their emerging fields as they sought to accommodate new perspectives on the world around them.

In this chapter, we will begin by briefly considering the relationship between classical and Hellenistic philosophy and the early Church. We will consider the impact of the fall of Rome on Western Europe, and will briefly touch on the scholarship of the Venerable Bede, before turning to the emergence of universities and the influence of Hellenic, Hellenistic, and Islamic scholarship on twelfth-century European scholars.[1] We will look in particular at the controversies regarding Aristotelianism in thirteenth-century Paris, and specifically at the work of Robert Grosseteste and the problem of the rainbow, in order to demonstrate the nature of the relationship between theology and natural philosophy during this period. Johannes Kepler's work on planetary movements forms another example. We will conclude by considering how the myth of the 'Dark Ages' could have ever been applied to a

Science, Religion, and the Human Future. Amanda Rees et al., Oxford University Press. © Amanda Rees, Franziska E. Kohlt, Tom McLeish, Charlotte Sleigh, David Wilkinson (2025).
DOI: 10.1093/9780191995316.003.0002

culture in which some scholars complained that there was even too much natural philosophy contained within theology. What medieval tools for a theology of science might we find useful in our contested present?

Doctrinal disputes and philosophical frameworks

Early Church fathers of the second and third century, such as Tertullian, Tatian, and Hippolytus of Rome, took issue with key aspects of classical Greek—that is to say, pagan—philosophy.[2] This is hardly surprising. As Christianity evolved from a minority faith to its position as the official creed of the Roman empire, positions needed to be both defined and defended. It was a period of tremendous doctrinal diversity. But it is important to realise that early Church apologetics was often undertaken on terms that ultimately derived from 'pagan' philosophy. Prior to their conversion, early Church figures would have been educated within a theoretical framework that was firmly based within classical philosophy. Their attitude and approach to argumentation and debate was formed within this framework, meaning that both for internal and external disputation, the analytical tools available to them originated within an approach to rational discourse that had been shaped by Hellenistic philosophy. Even where pagan thought conflicted with doctrine, early Church fathers had no choice but to continue to think like pagans in order to defend the faith. And where pagan thought did not conflict with doctrine, then what was the harm in engaging with it, or making use of it? There were many such areas, from geometry through rhetoric and natural history; none of these posed problems for scripture.

Some early theologians worried that too much curiosity about the natural world might distract people from engagement with the divine.[3] Cautioning against the lure of Greek philosophy, the Church father Tertullian famously asked: 'What has Athens to do with Jerusalem?' Yet at the same time, scriptural knowledge was infused with references to the natural world which the Greeks had studied. Augustine of Hippo (354–430) concluded that curiosity was not necessarily idle or disruptive but could be a form of worship in and of itself. Key passages of scripture, from the Psalms to the book of Job, focus the reader's attention on the world around them, requiring them to regard engagement with God's creation as engagement with the divine.[4] After all, if one did not understand the regularity of the divinely created natural order, how could one appreciate miraculous intervention within it? Augustine, in both his *Confessions* and the *Literal Meaning of Genesis*, did much to create a privileged place for natural philosophy within theology by stressing its utilitarian purpose: a Christian should recognise and use truth wherever it was found. In this way, natural philosophy would become the 'handmaiden' of theology—there to serve its mistress as well as to be protected by her.

One of Augustine's most important works, *The City of God*, was written in shocked response to the sack of Rome by the Visigoths in the early fifth century. Some Romans interpreted that event as divine punishment for abandoning their old gods for the new Christian faith. Augustine's work, covering everything from theodicy to eschatology, and including natural theology and the history of creation, aimed to refute this and to reassure his readers of God's role on Earth. But the fall of Rome also heralded a serious shift in the fortunes of both natural philosophy and theology within Western Europe. Since most cultured Roman citizens were fluent in both Latin and Greek, many key works of Hellenic philosophy had never been translated into Latin, including much of Aristotle's oeuvre. After the fall of Rome, few if any literate Western Europeans were capable of reading Greek. As a result, Western Europe lost access to many of these philosophical resources. Fortunately, the eastern half of the Roman empire, now centred on Constantinople, had retained its linguistic flexibility and was—with the rise of Islamic culture—shortly to see that linguistic capacity expand significantly.

By the time of the Umayyad caliphate in the late seventh century, there were already programmes underway to translate Greek natural philosophy into Arabic. These programmes accelerated significantly under the Abbasids, especially in the reign of al-Ma'mun, in the early ninth century. This work went far beyond mere translation from one language to another. The people involved had to create a new lexicon, a vocabulary through which Arabic scholars could engage with Greek natural philosophy. Checking hand-written texts from different copyists against one another, they reached for a consistent expression of ideas that would also accurately reflect external reality. This 'translation movement' was one of the great sustained periods of collective human creativity, developing new and original ways of thinking about the world which resulted in significant contributions to mathematics, optics, physics, medicine, and astronomy, among other areas. By the eleventh century, the philosophy that had originated with the Greeks had become thoroughly enculturated and naturalised within Islamic culture, as demonstrated by the widespread popularity of the *Treatise of the Brothers of Sincerity*, which popularised a Neoplatonic version of Aristotelianism.[5]

Ibn Sīnā, who lived during the late tenth and early eleventh centuries, wrote on a vast array of subjects that included alchemy, astronomy, geography, geology, psychology, logic, theology, poetry, and medicine. Avicenna, as he became known in the West, examined methods of natural philosophical enquiry, developing the 'Proof of the Truthful' to show the necessity of God, while at the same time developing a theory of motion that (unlike Aristotle's) made sense of the movement of an object under propulsion.[6] Meanwhile, the scholar Ibn Rushd dealt with medicine and optics—particularly the structure and function of the eye—as well as with jurisprudence. His contributions demonstrated the value of philosophy in and to Islamic theology, stressing the importance of interpretation when it came to dealing with divinely inspired writing. From his perspective, scripture did not need to, and sometimes should not, be understood literally. Explicitly defending natural philosophy from challenge by scholars such as Al-Ghazâlî (who himself used logical methods to attack philosophers), Ibn Rushd argued that reason cannot contradict revelation because truth cannot contradict truth. If there is apparent contradiction, he maintained, then the solution is for the meaning of scripture to be considered more carefully by those best qualified so to do. In Western Europe, Ibn Rushd later became known as Averroes, or 'The Commentator', because of his extensive re-working of Aristotle. In the tenth and eleventh centuries, however, as we have previously noted, European access to Hellenic philosophy remained extremely limited.

Scholars in the West at this time were reasonably familiar with Plato: Cicero (106–43 BCE) had begun the translation of works such as the *Timaeus* from Greek to Latin, with the result that the influence of Platonism can be seen in the work of writers such as St Augustine. Aristotle, on the other hand, had been largely lost, as had the mathematical and astronomical concepts and techniques created by Euclid and Ptolemy. The sixth-century Roman philosopher and Christian scholar Boethius, born just after the Western empire's fall, had done his best to preserve the classical tradition by translating as much as possible of Aristotle from Greek to Latin, but was only able to achieve a limited amount before being executed by Theoderic, King of the Ostrogoths. The translations that Boethius completed before his death, especially Aristotle's work on logic, represented a significant proportion of the theological and philosophical scholarship available to the West before the eleventh century. To these were added his own commentaries on Aristotle's work and the introductions which he wrote on music, arithmetic, and astronomy, as well as his *Consolations of Philosophy*.[7] Together, these writings provided churchmen such as Gerbert of

Aurillac (Pope Sylvester II), St Anslem of Canterbury, and the Venerable Bede with creative inspiration and analytical tools. Notably, however, the absence of access to the Hellenic corpus of astronomical philosophy did not prevent medieval scholars or laity from keeping a keen eye on astronomical events, or from developing mathematical theologies of time in the tradition of *computus*.

Computus, bureaucracy, and empire

Charlemagne's political unification of much of Western Europe in the late eighth century did not outlast his death. However, the conversion to Christianity which he enforced on his defeated opponents did create a unified and enduring religious culture for the continent. Monastic and cathedral schools flourished within it and went on to have enduring influence over its culture. Many students at these schools would go on to become clerics, but lawyers, doctors, and merchants also began their careers within their walls; Charlemagne's empire needed literate and learned administrators. Charlemagne's efforts to revitalise learning at his court and beyond brought scholar-clerics such as Alcuin of York to revive the liberal arts tradition, based around the trivium (grammar, logic, and rhetoric) and quadrivium (arithmetic, music, astronomy, and geometry). The emperor's engagement with natural philosophy can be clearly seen in a 798 exchange of letters with Alcuin; the two men discussed the retrograde movement of the planet Mars, and even, in Alcuin's case, posited for the sake of argument that the universe is sun-centred.[8]

Arithmetic and astronomy were critically important to one essential part of clerical education: the calculation of the date of Easter. This problem, expressed in the procedures known as *computus*, had been a matter of debate from the first Council of Nicaea in 325 to the Synod of Whitby in 664, with different parts of the Church holding strong and divergent views on the matter. Calculating the date on which Christians should celebrate Easter depended on resolving a number of factors. These included the date of the spring equinox and of its associated full moon, as well as the date (in early calculations) of the Jewish festival of Passover. The day also had to coincide with a Sunday. All this was complicated by the incompatibility of the solar and lunar calendars. Full moons happen on the same date every eight or eleven years, whereas a given solar date will fall on the same day of the week every twenty-eight years. Meanwhile, the solar year itself is slightly more than 365 days in length, and the moon takes slightly less than a calendar month to orbit the Earth. Thus, calculating the date of Easter was, to put it mildly, complicated. Monks and clerics used sophisticated astronomical and mathematical skills to create cycles and tables that successfully enabled the date of Easter to be determined years in advance.

While these debates depended on both astronomical observation and arithmetic expertise, they were about far more than just a technical process of choosing a date to celebrate Easter. On the contrary, these readings and calculations were an essential means to a sacred end. These were efforts to identify the moment in time at which the crucifixion and resurrection had taken place: that singular moment in which humanity was redeemed. Calculated correctly, it was also the moment at which the whole community of Christian faith would come together in unity to give thanks. Teachers in the schools had to be able to explain to their students the principles that underpinned these calculations, as well as their historical context. In this way, familiarity with concepts such as the equinox, eclipses, leap years, and planetary motion through the zodiac became an indispensable clerical skill. It was complemented by the use of astronomical observation to identify appropriate times for other festivals and saint days—in other words, to provide the calendar in which the life of the Church could take shape.[9]

Scholars associated with these medieval schools produced impressive and influential work. Anselm of Canterbury, for example, showed in his *Proslogion* how one could use logic to demonstrate the existence of God. The proof was not intended to sway a sceptic, since it begins from the assumption that God exists. Instead, Anselm was providing a service for those members of the faithful who wanted to see the rational grounds for their belief. Scholars also made good use of what access to Aristotle they had. One key use of Aristotelian philosophy was its relevance to discussions about transubstantiation and the Eucharist. Aristotle's theory of matter distinguished between accidental and substantive properties. Substantive properties were essential (a cat must be an animal) but accidental properties were variable (a cat can be black, or have three legs, or no tail, and still be a cat). In relation to the Eucharist, Anselm's teacher, Lanfranc, pointed out that the accidental properties of the host—its apparent 'breadness'—could be retained even as its substantive qualities were transformed into the body of Christ.

Aristotle's work also spoke into theological debates on the nature of universals (classes of quality in the world). Were such categories real, with an existence independent of human perception, or were they just constructed out of human names (nominalism)?[10] Key figures in these debates—Roscelin of Compiegne and Peter Abelard—would be accused of heresy for applying the doctrine of nominalism to the Trinity, although in Abelard's case at least, problems seem to have stemmed as much from arrogance and hypocrisy as from heretical belief.

All these scholars, however, were working with a more limited set of intellectual and methodological tools than those of their Islamic peers. This was becoming more obvious by the late eleventh century, as the cities of Toledo and Jerusalem, and the island of Sicily, were forcibly brought back under Christian control through the Reconquista and the Crusades. Spain, especially, provided Christian scholars with a trove of Islamic scholarship on astronomy and medicine, while Adelard of Bath travelled to Sicily to gather mathematical works, resulting in the first complete translation to Latin of Euclid's *Elements*. Gerald of Cremona translated Ptolemy's *Almagest* from the Arabic, as well as Aristotle's philosophy with commentaries by Ibn Rushd. By the late twelfth century, the scholars of Western Europe were gaining access to a vast quantity of previously unfamiliar Islamic-Hellenic natural philosophy, some of it difficult to reconcile with established Church doctrine.[11]

Instituting natural philosophy

At the same time, a new kind of organisation was emerging in Europe. Charlemagne had needed literate scholars to help run the administration of his empire in the eighth century; by the twelfth century, rulers felt that need even more strongly. Governments that had been relatively small and based primarily on interpersonal relationships were in the process of transforming into much larger structures. As emergent states tested the boundaries of their power over their subjects, as well as their relationships with one another, the value of organisation and bureaucracy grew. All this was rooted in the learning and reason of the Church. Legal scholars, working on the recovery and study of Roman law, were particularly significant, and a new kind of institution arose to house them: the university.[12]

Unlike monastic or cathedral schools, the new universities were not under the control of an ecclesiastical leader, such as an abbot or bishop. Instead, they were self-governing guilds or corporations, legally independent under charters issued by local and national rulers. Bologna was the first of these, followed by Paris in 1200, Padua in 1222, and Oxford at around the same time. By the end of the fifteenth century, around sixty universities were in existence in Europe. Along with them came the need to organise knowledge within curricula from which students could be taught. Programmes of learning were developed, based on key texts whose study would produce individuals with the skills for the expanding ranks of literate careers. Universities were producing professionals as much as they were producing and protecting knowledge.[13]

At the core of the new institutional curricula were the seven liberal arts, divided into the *trivium* (the verbal arts of grammar, logic, and rhetoric) and the *quadrivium* (the mathematical arts of arithmetic, geometry, music, and astronomy). These subjects taught basic thinking skills, and as such were an essential prelude to any advanced work; indeed, many students did not proceed beyond their study. But anyone who wanted to progress to the 'higher' faculties of law, medicine, or theology had first to master the basics of natural philosophy. In practice, what this meant was that while all theologians would have had a thorough training in natural philosophy, becoming a theologian took many years of study beyond that basic philosophical grounding.

Because theologians needed to have the skills of natural philosophy, the scholars of Western Europe had to reach an accommodation with the new Islamic-Hellenic learning that was arriving from the continent's peripheries. In many cases, this was not problematic. William of Conches, for example, was able to reconcile the book of Genesis with Plato's *Timaeus*, not least by explicitly acknowledging (as had many medieval scholars in Europe and Asia before him) that there was no need to take the Bible literally.[14] Divinely inspired, it had been written so as to be understood by ordinary people: scholars would naturally interpret its allegories and metaphors in a more sophisticated manner. Aristotle's work, however, was a different matter. The general naturalistic approach which he adopted could be dealt with through the distinction between primary and secondary causes: accepting God as the primary cause did not prevent anyone investigating secondary causes of action (what we might now call physics). But there were also a number of points at which Aristotle's work seemed to flatly contradict doctrine, particularly with reference to the nature of the soul, the eternity of the world, and the existence of any constraint on divine action. These issues came to a head at the University of Paris in the thirteenth century.

The debates of Paris

The conflict between the bishops of Paris and the city's arts faculty between 1210 and 1277 has been frequently adduced to demonstrate that science and religion have always been at war. However, it is important to consider the context of what was happening in Paris over this period. This was not an example of the Church banning science. It was instead a series of debates between theologians, churchmen, and philosophers that took place over a number of decades in the context of a changing intellectual and political landscape. These debates also focused on the work of the philosopher Aristotle whose ideas had, by the end of the century, become an essential element in Christian theological doctrine.

In 1210, Peter of Nemours, then Bishop of Paris, became aware of the fact that doctrines of pantheism and materialism (the divinity of matter) were being discussed and preached in and around the university by followers of Amalric of Bena. Amalric, who had been an influential teacher at the university before his death, seemed to have derived these doctrines from his reading of Aristotle. The bishop's response was swift: Amalric's disciples were ordered to renounce him, and those who refused were handed over to the state for execution. The bishop and his synod banned the books of Aristotle on the spot. But crucially, this ban only applied to the arts faculty at Paris, and not to other universities in France. Even this limited ban was relatively swiftly watered down by Pope Gregory IX, who directed that the books could be taught freely again once they had been edited to remove any specific problems. It is not at all clear if anyone actually tried to do this, but the fact that the ban was renewed in 1228 suggests that if it was done, it was not done effectively. Even in its renewal, though, the ban was still restricted to the University of Paris—at least, that is, until the university at Toulouse began to advertise itself as a place where Aristotle could be freely taught. At this, the Pope promptly extended the ban to Toulouse as well. Nevertheless, it is clear that, whatever the official position of the Church may have been, Aristotle remained central to the work of both the philosophers and the theologians at the university.[15]

By the second half of the thirteenth century, philosophers such as Siger of Brabant were largely ignoring Christian doctrine as they explored the thinking of Aristotle, Averroes, and others. In 1277, Pope John XXI asked the Bishop of Paris, Étienne Tempier, to investigate heresy in the city. In response, the bishop published a list of 219 condemned propositions relating to Aristotelian texts that he deemed to be irreconcilable with Christian belief. Five years previously, new members of the arts faculty at Paris had been required to swear that they would avoid dealing with theology in their philosophical discussions, and that if theological matters did arise during the course of debate, they would ensure that the debates were resolved in favour of doctrine. While this might look like a sustained attack on Aristotelian thinking from the Church, it can also be read as a demonstration of the extent to which churchmen continued to study, engage with, and develop Aristotle's work. Certainly, some of the bishops of Paris were willing to attack 'pagan' philosophy, but not all of them did so by any means. William of Auvergne, who held the position from 1228 to 1249, is a good example. Like Augustine of Hippo, he was a powerful advocate of the importance and necessity of reason in theological debate. In his *Magisterium Divinale et Sapientiale*, he engaged closely with Aristotle, seeking ways of reconciling him with Christian doctrine.

Of even greater significance throughout this period was the work of Albertus Magnus and of his student Thomas Aquinas. Between them, they simultaneously Christianised Aristotle and Aristotealianised Christianity. Both men were linked to the Dominican order of friars (founded in the early thirteenth century) and their works systematically examined and expounded on the writings of Aristotle, identifying problems and showing the faithful how they could make sense of them from within Christianity. Even more significantly, Aquinas' *Summa Theologica*, an unfinished work that continues to resonate through Western culture at the present day, used Aristotle's concepts and methods to interrogate and investigate theology itself. Bishop Tempier might have included some of Aquinas' propositions on the 1277 list of condemnations—but this did not prevent Pope John XXII from canonising St Thomas in 1323.

Fundamentally, the natural philosophy of Aristotle, Euclid, Ptolemy, Averroes, Avicenna, and others was just too useful to be disregarded for long. Perhaps, putting difficult philosophers on the curriculum was simply a useful gatekeeping device for advanced professions. Not everyone would manage to master the material: as a result, the value of those who could command its expertise was increased. But more significantly, these works fed a very real hunger amongst scholars for intellectual tools that could help them apprehend and engage with the world around them, not least because by so doing, they grew closer to God. As St Augustine had shown, studying the order of the world that God had created was itself a form of worship; as the example of *computus* demonstrates, the order of nature could and would determine the order of worship. Even more importantly, far from being inherently heretical, natural philosophy would give the Church the tools with which to defend the faith and to convert non-believers. This task was all the more urgent for those, like the Franciscan Roger Bacon, who believed that the end of the world was near. It is possible that one reason for the lifting of the Paris ban was pressure from the Dominican order, who were aware of how useful Aristotle would be in their debates with the heretic Cathars.[16]

Enlightenment and the rainbow

Medieval scholars saw a profound connection between natural philosophy and theology, as the life and work of Robert Grosseteste shows. Bishop of Lincoln from 1235 until his death in 1253, with connections to Paris as well as to the Oxford Franciscans, Grosseteste wrote in highly mathematical ways about light, colour, sound, and the heavens. He drew on the earlier Islamic transmission of, and commentaries on, Aristotle. He went on to develop many topics in astronomy, mechanics, and, above all, optics, well beyond the Islamic legacy. Grosseteste was, for example, the first to identify the phenomenon of refraction in relation to the rainbow. His understanding of the point and purpose of natural philosophy sat within an overarching Christian narrative of creation, fall, and eventual redemption.[17]

Like other scholars, Grosseteste linked scriptural truth with the truth of the natural world. In his commentaries on the Psalms, he noted that if the Bible uses natural objects to illustrate and illuminate its meaning, then it is incumbent on the faithful to discover as much as they possibly can about those objects, in order that the fullest meaning and implications of scripture can be conveyed. The metaphor of 'seeing' as 'understanding' was central to Grosseteste's understanding of theology and natural philosophy. In his *Commentary on the Posterior Analytics* (of Aristotle), he showed how redemption could come through reasoning. Humanity's 'higher' powers relating to reasoning and the spirit had been blunted by the Fall: the only hope of recovering them was found through the deployment of the 'lower' senses of sight, sound, and so on in the contemplation of God's creation. Making sense of what these senses reveal required the exercise of both the scientific imagination and mathematical reasoning to 'see' the connections and causes between events and processes in the world. A higher power or interpretive understanding was thereby awakened (Grosseteste called it *sollertia*) through which the inner workings that lay behind external appearance were revealed to the close observer: 'Aha! I *see*' ... Grosseteste was neither the first nor the last to speak in these terms. Macrina, the sister of Gregory of Nyssa, described her perception of the phases of the moon as a means to demonstrate the activity of the soul. We 'see' circles and crescents waxing and waning even though we know that what we are watching is a sphere passing through different angles of illuminating sunlight. The mind, in Macrina's description, guides the eye to understand the deeper meaning of what is apparently visible. Macrina makes a theological parallel with the hymn to wisdom in Job 28: not even the clear-sighted falcon can see the world with God's understanding. Only humanity is invited to share it, as a step towards redemption. Grosseteste, describing Macrina's thoughts, entertains an early conception of the material way in which humankind is, literally, Earthed into creation.[18] Enlightenment and light connect humans with corporeal nature and with the heavens above.

If science was for Grosseteste the source of human redemption, then the rainbow was the signal of regeneration. As the symbol of God's covenant with Noah and of his commitment to creation, the rainbow had been the focus of philosophical attention since Aristotle. Explaining the rainbow depended on an understanding of both optics and the nature of colour. In a largely pre-literate age, colour was a key ecclesiastical sign for the laity to read.[19] Grosseteste's work on colour (*De Colore*), light (*De Luce*), and the rainbow (*De Iride*) uses the rainbow—symbol of God's promise to the world—to explore and illuminate his theories of colour and light.[20] Grosseteste suggested that the physical cause of the rainbow was refraction in a cloud (not, as Aristotle thought, reflection), though it was not until early in the fourteenth century that the underlying process was identified and mathematically described. Inspired by Ibn al-Haytham's work on optics, Kamāl al-Dīn al-Fārisī, working in Baghdad, and the Dominican friar Theodoric of Freiberg, simultaneously used experimental methods to demonstrate how light interacted with raindrops to produce the rainbow. For Theodoric at least, this was at least partly inspired by God's question to Job: 'By what way is the light parted, which scattereth the east wind upon the earth?'. Unlike the fragmented light of rainbow, the inspiration and practice of theology and natural philosophy cannot be separated out for these early investigators.

A hidden light

The catalogue of scientific contributions made by medieval scholars is undeniable and impressive. The fourteenth-century Archbishop of Canterbury, Thomas Bradwardine, produced a demonstration that motion could be described mathematically. William Heytesbury developed a version of the mean speed theorem by mathematical reasoning. Nicole Oresme, Bishop of Lisieux, provided a mathematical proof of the mean speed theorem as well, developing a new theory of motion. Oresme also refuted most of the objections to a moving Earth before Copernicus, demonstrating (again) that medieval scholars saw no need to take the Bible literally. As late as the seventeenth century, Johannes Kepler was able to solve the problem of planetary motion precisely because he, like his medieval predecessors, believed that consistency was at the heart of God's creation of the cosmos. God meant the universe to be understood by human minds: underneath the complexity of circular orbits, epicycles, and eccentrics, therefore, must lie a simpler solution. This he found in ellipses and the insight that planetary velocity varies regularly according to distance from the sun. As Kepler put it: 'For a long time I wanted to be a theologian. Now, however, behold how through my effort God is being celebrated through astronomy'.[21]

The bans on discussing Aristotle in Paris may, as Pierre Duhem argues, actually have given impetus to the exploration of new scientific concepts and theories. The list of condemnations that Tempier drew up is incoherent and disorganised at best, but at their root is the insistence that God has untrammelled freedom of action: any Aristotelian claim that placed limits on that creative freedom was, therefore, heretical. By refusing to accept any constraint on God, the ban opened up imaginative and creative space within which scholars could contemplate alternative worlds and ways of being. Indeed, Nicholas of Cusa went so far as to place God, not the Earth, at the centre of the universe and to claim that life may be found in every region of solar and stellar space.[22]

Science and religion were not at war during the Middle Ages. It is true that individuals, and the institutions to which they belonged, frequently came into conflict with each other with respect to the authority accorded to theology or natural philosophy. Simultaneously, however, the institutions of the Church provided a secure place, both economically and intellectually, within which scholars could explore their understanding of the natural world, grounded in the role of a consistent Creator and the capacity of the mind to bring the soul closer to God. Medieval science, far from being an oxymoron, played an essential role in the events lauded by later historians as 'the Scientific Revolution'. Not only Newton, but Copernicus, Kepler, and even Galileo, stood on the shoulders of giants in order to be able to 'see' (in the sense that Robert Grosseteste would have understood) into the world around them.

The medieval achievements described in this chapter may surprise some readers. They have, indeed, been consistently downplayed and overlooked by scholars in the period between then and now. This occurred in part due to a commitment to the idea that the ancient world was the cradle of civilisation, and that nothing which happened between the fall of Rome and the Renaissance was relevant to the modern world. Later scholars, from the enlightenment onwards, were also anxious to stress the novelty of their own work. They systematically undermined medieval natural philosophy, designating as 'Dark Ages' the period that followed the fall of Rome. This picture of the European Middle Ages was further weaponised in the late nineteenth and early twentieth centuries as the story about the 'war' between science and religion began to gain traction. It is chastening to reflect on what the medieval approach might still offer us today, not only in terms of our understanding of science and theology, but perhaps with respect to a theology of science that places questions of purpose, ethics, and ultimate responsibility at the forefront of understanding. We will pick up on this theme in the third and final section of this book.

Further reading

Sonja Brentjes, *Teaching and Learning the Sciences in Islamicate Societies (800–1700)* (Turnhout: Brepols, 2018)

Seb Falk, *The Light Ages: A Medieval Journey of Discovery* (London: Allen Lane, 2020)

Rivka Feldhay and F. Jamil Ragep, eds, *Before Copernicus: The Cultures and Contexts of Scientific Learning in the Fifteenth Century* (Montreal: McGill-Queen's University Press, 2017)

Tom McLeish, *Faith and Wisdom in Science* (Oxford: Oxford University Press, 2016)

Tom McLeish, 'Lessons in the Distant Mirror of Medieval Physics', in *After Science and Religion: Fresh Perspectives from Philosophy and Theology*, edited by Peter Harrison and John Milbank (Cambridge: Cambridge University Press, 2022), 259–281

George Saliba, *Islamic Science and the Making of the European Renaissance* (Cambridge, MA: MIT Press, 2011)

Notes

1. 'Hellenic' refers to any philosophy originating in Greece; 'Hellenistic' refers to the Greek-influenced world in the period from the death of Aristotle and up to the Roman conquest of Egypt in 30 BCE.

2. David C. Lindberg, 'Medieval Science', in *The History of Science and Religion in the Western Tradition* edited by Gary B. Ferngren, Edward J. Larson, and Darrel W. Amundsen (New York: Routledge, 2000).

3. P. G. Walsh, 'The Rights and Wrongs of Curiosity (Plutarch to Augustine)', *Greece & Rome* 35.1 (1988): 73–85, http://www.jstor.org/stable/643280.

4. David C. Lindberg and Michael H. Shank (eds.), *The Cambridge History of Science. Volume 2: Medieval Science* (Cambridge: Cambridge University Press, 2013).

5. Jan P. Hogendijk and A. I. Sabra, *The Enterprise of Science in Islam: New Perspectives* (Cambridge MA: MIT Press, 2003).

6. Peter Adamson, *Ibn Sīnā (Avicenna): A Very Short Introduction* (Oxford: Oxford University Press, 2023).

7. Noel Harold Kaylor Jr. and Philip Edward Philips, *Boethius in the Middle Ages* (Leiden: Brill, 2012).

8. Tom McLeish and Mary Garrison, 'Reversals in Wartime: Alcuin and Charlemagne Discuss Retrograde Motion', *Interfaces: A Journal of Medieval European Literatures* 8 (2021): 14–53, https://doi.org/10.54103/interfaces-08-03.

9. Seb Falk, *The Light Ages: A Medieval Journey of Discovery* (London: Allen Lane, 2020).
10. Stephen P. Marrone, 'Medieval Philosophy in Context', in the *Cambridge Companion to Medieval Philosophy*, edited by Arthur McGrade (Cambridge: Cambridge University Press, 2003).
11. David C. Lindberg (ed.), *Science in the Middle Ages* (Chicago: University of Chicago Press, 1978).
12. Michael Shank, 'Schools and Universities in Medieval Latin Science', in *The Cambridge History of Science. Volume 2: Medieval Science*, edited by David C. Lindberg and Michael H. Shank (Cambridge: Cambridge University Press, 2013).
13. Toby Huff, *The Rise of Early Modern Science: Islam, China and the West* (Cambridge: Cambridge University Press, 1993).
14. Joan Cadden, 'Science and Rhetoric in the Middle Ages: The Natural Philosophy of William of Conches', *Journal of the History of Ideas* 56.1 (1995): 1–24, https://www.jstor.org/stable/2710004.
15. Edward Grant, *The History of Science and Religion in the Western Tradition* (London: Routledge, 2000).
16. Roger French and Andrew Cunningham, *Before Science: The Invention of the Friars' Natural Philosophy* (London: Routledge, 2016).
17. Tom McLeish, 'Lessons in the Distant Mirror of Medieval Physics' in *After Science and Religion: Fresh Perspectives from Philosophy and Theology*, edited by Peter Harrison and John Milbank (Cambridge: Cambridge University Press, 2022).
18. Tom McLeish, *Faith and Wisdom in Science* (Oxford: Oxford University Press, 2014).
19. Herman Pleij, *Colours Demonic and Divine: Shades of Meaning in the Middle Ages* (New York: Columbia University Press, 2004).
20. Hannah Smithson, Giles Gasper, and Tom McLeish, 'All the Colours of the Rainbow', *Nature Physics*, 10 (2014): 540–542, https://doi.org/10.1038/nphys3052.
21. C. M. Linton, *From Eudoxus to Einstein: A History of Mathematical Astronomy* (Cambridge: Cambridge University Press, 2004). Quoted in James Hannan, *God's Philosophers* (London: Icon, 2009), p. 291.
22. Michael Crowe, *The Extraterrestrial Life Debate: Antiquity to 1915* (Indiana: University of Notre Dame Press, 2008); David Wilkinson, *Science, Religion and the Search for Extraterrestrial Intelligence* (Oxford: Oxford University Press, 2017).

2

Enlightening imperialism

Movement apart, movement in concert

Between the late eighteenth and the early twentieth century, profound synergies between three globalising phenomena came together to create a world that was very different from that which had been experienced by medieval scholars. Scientists searched for natural laws that would apply as certainly in Bengaluru as they did in Birmingham or Bissau. Capitalists were hungry for new markets and new resources to exploit. Christians were anxious to bring the word of God to all of the world's peoples. These three trends combined in complex, and sometimes contradictory ways, in the empire-building activities of European (and European-settled) nations. By the mid-nineteenth century, even as the public sphere increasingly avoided discussions of personal faith, science continued to play a key role in missionary activities. Men of science (and they were usually men) participated in scientific expeditions that mapped routes and identified raw materials for military and merchant usage. The 'civilising mission' of faith and science together provided Europeans with a joint and rational justification for empire that sat comfortably alongside the drive for power and profit.

As this chapter will show, this was also a period marked by profound changes in the ways in which 'science' and 'religion' were understood. As Peter Harrison and others have demonstrated, they were no longer purely associated (as were *scientia* and *religio*) with the internal virtues possessed by an individual. Instead, they increasingly became reliant on the existence of external institutions, bodies of knowledge, and established practices. These institutions, their literatures and practices, began to diverge. Science became more visible amongst the middle-class public, while by the later nineteenth century, faith became a matter to practise in private. While faith continued to play a role in the personal life of some men of science, it was no longer necessarily a matter for public espousal.

The two spheres, however, found common cause when it came to the concept of 'progress'. Empire, Christianity, and science came together in

Science, Religion, and the Human Future. Amanda Rees et al., Oxford University Press. © Amanda Rees, Franziska E. Kohlt, Tom McLeish, Charlotte Sleigh, David Wilkinson (2025).
DOI: 10.1093/9780191995316.003.0003

a tangled skein that roped political power and profit together with deeply and genuinely held belief. Ideas about both cultural and biological evolution informed the European encounter with other societies and communities, and the information obtained about those communities—not least by missionaries—fed back to the narratives and analyses that European scholars were constructing about the nature of humanity. By the late nineteenth century, European men of science believed they had good reason to assert themselves as the representation of the apex of human development. As a headline in *The New York Herald* brayed at the turn of the century, 'Men of the Future' might even 'Become as Gods'.[1] This chapter will show how that happened and what this implies for the way that the relationship between science and religion is understood.

Academies, societies, and a new kind of public space

The seventeenth and eighteenth centuries in Europe have traditionally been described by historians as the 'Age of Enlightenment', a phrase that characterises a series of cultural, political, philosophical, and economic transformations in ideas about God, nature, and humanity. All these, so the story goes, were now understood via reason and rationality rather than faith. These ideas were not just being discussed in universities or ecclesiastical institutions but were debated in new kinds of public spaces: the lodges, coffeehouses, and salons, as well as in pamphlets, books, and periodicals. At the same time, the concept of ruling by divine right and the role of churches in supporting these claims to political and economic power were shifting in the aftermath of the post-Reformation wars that had dominated European politics since the early sixteenth century. Scholars such as John Locke, Thomas Jefferson, and Thomas Paine supported calls for religious toleration, which, while frequently contested, shaded into wider demands for liberty, freedom, and the rights of the individual.[2]

Demands to put the governance of social life on a 'rational' basis also had implications for the ways in which European intellectuals understood the role that reason had played in their own histories. In justifying their calls, scholars identified themselves with the classical philosophers of Greece and Rome, thinkers who had first explored the capacity of reason to structure society. This vision of social life, argued Voltaire and Rousseau, should be the model for human interaction, not the faith-and-doctrine-based constraints associated with the Church. It was, claimed these scholars, the ancient world that was the true cradle of civilisation: after the fall of Rome, nothing of enduring human value had been achieved in the dark ages of superstition that followed. This conclusion could only be achieved by ignoring the accomplished scholarship of medieval writers, and indeed, sometimes by active destruction of their work by losing, selling, or reusing manuscripts.[3] Rationality, and a reasoned approach to the world, identified eighteenth-century Europeans as the true heirs of the ancient world, who would carry forward their philosophical and scientific legacy.

Learned societies such as the Royal Society of London (established 1660) or the Académie des sciences (1666) were founded to promote what the English group of natural philosophers identified as the new 'Physico-Mathematicall Experimental Learning'. In the French case, they were also to act as an integral part of the state apparatus: by the eighteenth century, their English counterparts too were regularly asked to provide scientific advice to the government on significant matters of the day. Literary and philosophical societies were being formed in other cities and towns. Influenced by the empirical and sceptical approaches articulated by thinkers such as Francis Bacon, Descartes, and others, they hosted discussions and demonstrations from respected national and local speakers. It was to the Literary and Philosophical Society of Newcastle, for example, that George Stephenson demonstrated his miners' lamp in 1815. Members of the Society, as well as miners from Killingworth Colliery, would later testify in Stephenson's defence during his controversy with Humphry Davy over who should get credit for the safety lamp.

The existence of these regional and local societies demonstrates that these new approaches and developments did not just appeal to social elites but drew broad interest from the general public. From Bernard de Fontenelle's *Conversations on the Plurality of Worlds* (1686), which used French rather than Latin to give an account of Copernican astronomy (and incidentally, was another early exploration of the possibility of extraterrestrial life), to Emilie du Châtelet's popularisations of Newtonianism, the literate public gained familiarity with the new sciences. Indicative of this trend, the Royal Institution opened its doors to Londoners in 1799 for the express purpose of diffusing knowledge, making information about invention, experimentation, and the application of science to ordinary life available to ordinary people. Attendance at the lectures—especially those given by Humphry Davy—was high, including a significant proportion of women.[4] Samuel Taylor Coleridge is reputed to have attended Davy's lectures in order to increase the number of metaphors he could use in his own poetry.[5]

Davy's popular lectures sometimes touched on religious themes. But explicit calls on God when accounting for events in the natural world became much less frequent during the nineteenth century, irrespective of the debates around Darwinian evolution. As Ronald Numbers and other historians have shown, by the end of the eighteenth century, materialist explanations had become far more common in framing philosophical discussion; by the mid-nineteenth century, God's place in scientific debate had reduced significantly.[6] This did not mean that science, or even natural philosophy, had become secularised. The debates surrounding Darwinian evolution in the latter half of the century made this very clear. At an individual level, many men of science continued to espouse and expound various formulations of deism (belief in a non-interventionist God) as an appropriate position of faith. From their perspective, this kind of theology could happily co-exist within a naturalistic worldview. Christian faith continued to be a factor in admittance to the universities of Oxford, Cambridge, and Durham until the Universities Tests Act of 1871. But by and large, by the end of the century, religion was no longer an explanatory factor available to be used by men of science when accounting for the natural world.

Identifying the 'men of science'

Famously, William Whewell invented the term 'scientist' in 1833 in response to the call for a term to describe students of the material world. It certainly had the benefit of not immediately excluding women: in fact, one of its first usages in print was in Whewell's review of the astronomer Mary Somerville's *On the Connexion of the Physical Sciences* (1834). But it was not a term that gained much popularity until much later on. For most of the nineteenth century, the terms 'men of science', 'cultivators of science', 'scientific workers'—or even 'votaries of science'—were much more common.[7] The meanings attached to these terms shifted considerably over time in relation to changing concepts of expertise and authority over the natural world. Both had implications for the continuing role of faith in these matters.

When the Royal Society was founded in 1660, it was, above all, a society of and for gentlemen, a significant number of whom also followed one of the learned professions: medicine, law, or theology. Because they were gentlemen, the Society's members could be trusted to tell the truth about what they saw, guaranteeing the authority and significance of its demonstrations and debates. As gentlemen, they also belonged to the ruling class:

their knowledge could thus readily be put to public service, tied into a web of patronage and influence that permeated the British state and society by the late eighteenth century. This was evidenced by the Society's move to Somerset House in the physical and political centre of London in 1780. By the early nineteenth century, however, ideas about authority and expertise were in flux, particularly with regard to the question of who one could or should trust: gentleman-researchers, or specialised experts?

The tussle over control of the Royal Society after the death of Joseph Banks in 1820 is an excellent example of this. Banks had been the naturalist on HMS *Endeavour*, Captain James Cook's first voyage to Australia and New Zealand, as well as an advisor and supporter of the evangelical London Missionary Society (LMS), itself inspired by Cook's expeditions. Banks had returned from the antipodes to immediate fame at home and in 1778 was elected as President of the Royal Society, a position that he held for the next forty years. But by the time of his death, his dominance over English science was being challenged by a new generation of researchers, who were establishing themselves in new societies—not only the Royal Institution or the Linnean Society (1788), but groups such as the new Geological Society of London (1807) and the Astronomical Society of London (1820). These men did include the kind of gentleman-scientist that Banks would have recognised: Humphry Davy, for example, was one of the founding members of the Geological Society. Their ranks also included nonconformists, abolitionists, and people whose fortunes had been made recently from trade. Their understanding of how to do science was very different from their predecessors'; Banks' idea of the cultured and wide-ranging 'gentleman of science' conflicted sharply with the more specialist approach that was characteristic of the rising generation.

In many ways, the founding of the Geological Society represents the emergence of what might now be considered the 'disciplined expert': their authority and their trustworthiness depended on their detailed knowledge of a particular area of learning. Knowledge was not to be simply a gentleman's accomplishment or a pastime but instead had to be pursued with a disciplined body and mind—a vocation which required hard labour and sacrifice. These men understood themselves as public-spirited, devoting themselves to the propagation of useful knowledge. They sincerely believed that improving the nation's access to better food, water, transport, housing, and communication were activities deserving of financial support and public appreciation. At the beginning of the nineteenth century, it was the social background of the individual practitioner that made his scientific activities and advice trustworthy and respectable; by its end, it was the practice and discipline required by science that made men of science worthy of respect and trust. Supporters of the old guard succeeded in ensuring the eventual election of the royal Duke of Sussex as president of the Royal Society to succeed Banks; nevertheless, the new approach could be seen in the growth of support for his opponent, the astronomer John Herschel.

The accomplished feats of engineering (canals, bridges, railways, docks, and tunnels) that characterised the Victorian period in Britain depended upon the development of disciplined habits of self-governance. These habits were echoed in the laboratory; experimental success required a similar capacity to make repeated close observations, as well as precise and careful measurement. Self-discipline was essential to both science and engineering. This was to be a key part of the identity of the disinterested 'men of science'—the tools and techniques through which knowledge was created were to be transparent to all, since they had nothing to hide in their search for truth for truth's sake. As John Herschel argued in a book published in 1831, knowledge 'should be divested, as far as possible, of artificial difficulties, and stripped of all such technicalities as tend to place it in the light of a craft and a mystery'.[8]

Such a conscious process of identity formation, a collective self-awareness, emerged amongst members of the many scientific societies. These men (and some women) read the same journals, attended the same meetings, and sought to see their achievements recognised by their fellows as well as society at large. The formation of the British Association for the Advancement of Science (BAAS) in 1831 brought together all of Britain's Literary and Philosophical Societies under one banner, signalling the emergence and awareness of new collective and disciplinary identities. Individual sections of the BAAS (Geology, Maths and Physical Science, Engineering, Agriculture) organised themselves, but each still operated under the overall umbrella of 'Science'. Annual meetings of the BAAS received detailed coverage in the national, regional, and local press. These were not rarefied exchanges between specialists; instead, they were treated as being of direct interest and relevance to the public at large. As with the Royal Institution, the assumption was that scientific knowledge was applicable to ordinary life; it could and should inform public decision-making and behaviour.

Debates at the BAAS meeting of 1861 demonstrated this public spirit in a manner that intersected with the continuing spiritual life and faith of men of science. For some years, issues of quality control had been causing problems for the telegraph industry in general and for the Atlantic Telegraph Company in particular. As a result of the discussions at that meeting, the BAAS established a committee to establish standards in electrical measurement that would have a global application. A key figure on this committee was James Clerk Maxwell, who was appointed as the first professor of experimental physics at Cambridge University in 1870. His reasons for accepting the position and the principles under which he designed and built the Cavendish Laboratory are a fascinating example of the mutual interweaving of science, religion, discipline, and measurement. In essence, Maxwell worried that the kind of work that was needed for success in physical investigation would seem to be too laborious for Cambridge students: that it would appear too similar to factory or manual labour for their parents to permit. Nothing, he argued, could be further from the truth. Christian tradition supported and valued physical engagement with the world. Measurement was not drudgery, but itself represented evidence of a commitment to a higher calling. God's creation of the universe depended on precision and uniformity: standardisation and strict accuracy were the closest that humans could come to following in God's footsteps and furthering the work of the Creator on Earth.[9]

The drive towards self-improvement that underpinned the founding of the Literary and Philosophical Societies was also linked to the growth of evangelicalism in England in the first half of the nineteenth century. Both found expression in Samuel Smiles' best-selling book *Self-Help: With Illustrations of Character and Conduct*, which appeared in the same year as Darwin's *Origin of Species*. Smiles advised his readers to be responsible, self-disciplined, thrifty, as well as (by the second edition) perseverant. What is particularly interesting is that he offered his readers engineers as role models. He told his audience that reading about the lives of self-made men such as Thomas Telford or James Watt was almost equivalent to reading the Gospels themselves. These exemplary characters had, using their own resources, practised their skill and capacity for invention. They regularly confronted and overcame what to another might have seemed insurmountable difficulties; their lives taught 'high living, high thinking, and energetic action for their own and the world's good'.[10] The public spirit of engineers and their capacity to sacrifice themselves to create a better world brought them closer to the grace of God. They echoed and exemplified evangelicalism's emphasis on hard work and self-discipline in Christian life.[11]

Science and governance

If engineers could bring people closer to God, then it should not be surprising that missionaries had an equally important role to play in both conducting scientific research and communicating scientific, technological, and medical knowledge. One of the most famous of all British missionaries, David Livingstone, encapsulates this phenomenon. In 1838, having already begun to train as a doctor, he joined the London Missionary Society (LMS), which Joseph Banks had helped set up forty years earlier. The missionaries supported his medical education as well as his training for the ministry. Once qualified, he was sent to Southern Africa in 1840 to preach the Gospel and, by helping to establish alternative trade routes for European goods, to discourage the slave trade. He travelled with galvanic batteries and a magic lantern in his luggage, in order that he could more effectively engage the imagination and the senses of his audiences, showing them slides of Bible stories alongside his preaching. He took astronomical readings, mapped the areas over which he travelled as he set up mission stations, and made close observations of natural history. By the 1850s, his connection with the Missionary Society was becoming less important to him than his connections with the Royal Geographical Society, the Royal Society, and the Foreign Office. Even as he parted ways with the LMS, however, he maintained his link to mission work. The Anglican Universities Mission to Central Africa, for example, was founded in 1860 in response to his personal appeal, drawing in the resources of Oxford, Cambridge, Durham, and Dublin.

Livingstone was by no means the only individual whose mission work went hand in hand with their scientific contributions, nor the only missionary to use science and technology as a means of supporting their preaching, as Sujit Sivasundaram's study of missionary science during this period makes clear.[12] At a basic level, missionaries had to engage with linguistic science to communicate with local people, particularly if they were to translate the Bible into indigenous languages. Natural history was essential: it helped them identify local resources and threats and formed the subject matter for the education of locals. It is not coincidental that Livingstone's *Missionary Travels and Researches in South Africa* (1857) bore the image of a tsetse fly on its title page. Botanical knowledge was important for another reason: missionaries sent samples, seeds, and plants home for investigation and propagation, as Pere Armand David did for the Muséum national d'Histoire naturelle while travelling in China and Tibet from the

1860s. Suzanne Aubert, working in New Zealand, learned from her Māori congregation the use of local herbs, recording their knowledge and monetising it to support her mission. Missionaries also had a role to play in bringing non-native plants and animals as part of efforts to 'improve' foreign landscapes to European eyes.

There was nothing new about this; Jesuit and Anglican missionaries in China and India had, in earlier centuries, used the telescopes and mathematics of European science to demonstrate the superior capacity of European science to predict astronomical events. If European science was to be preferred, they argued, then so too should European religion. The sheer number of missionaries that were sent out from Europe and America in the nineteenth century means that generalising about their activities and motivations is not easy; it is equally difficult to draw a clear line between their scientific and religious endeavours. This does not mean that there was no conflict between the two. As Livingstone found in his correspondence with the LMS, it was possible for authorities at home to feel that their missionaries were spending too much time on science and not enough on preaching. However, the analysis of local flora, fauna, and minerals found productive synergies within and alongside theologies of nature.

The information that missionaries sent home did not only concern local plants and animals. Anthropology was a central part of mission science, and missionaries contributed essential information to the European understanding of evolution, social life, and humanity. Central to these debates was the Enlightenment concept of 'progress', often understood as a combination of scientific and social phenomena. A good example here is the founder of positivist philosophy, Auguste Comte, and his 'Law of Three Stages'.[13] Comte argued that the first stage of social development was 'theological', where events were believed to be divinely caused. The second was 'metaphysical', where direct divine intervention was replaced by belief in a more abstract set of causes, while the third was the 'positivist' or 'scientific' stage, where events could be explained through wholly naturalistic means, based on observation, experiment, and the scientific method. Comte noted that development through these three stages could also be seen in individual development, with the theological stage being linked to childhood, and the scientific stage representing the maturity and rationality of adult responsibility. These ideas were critically important to the interpretation of the anthropological observations made by Europeans in Africa, Asia, and elsewhere—and equally significant to the treatment of the role of religion within Europe itself.

Evolving ideas

Ideas about socio-cultural evolution long predate the concept of evolution through natural selection that is associated with Charles Darwin and Alfred Russel Wallace. The observation that social and biological phenomena change over time stretches back at least as far as the classical philosophers. At the turn of the nineteenth century, both Adam Smith and Hegel had put forward different schemas of economic evolution, while Jean-Baptiste Lamarck had proposed his theory of biological evolution through transmutation. During the nineteenth century, scientists combined information about the societies of colonised peoples with discoveries about human prehistory. They created linear, ladder-like models of human bio-social evolution that provided a scientific justification both for European imperialism and for the secularisation of European states and societies.

Fundamentally, the social philosophers of this period were trying to understand the differences between the state they understood as 'civilisation', such as that seen in England, France, the United States, Germany, and so on, and the state that they understood as 'primitive'. Primitive societies existed both in the European past and amongst the present-day peoples living under colonial rule. How, asked philosophers and scientists, might primitive communities progress to become civilised societies? The Great Exhibition of 1851 in London demonstrated this conundrum: by bringing together examples of the socio-technological achievements of all the nations of the world, it seemed to be very clear that some nations had progressed much further than others in terms of their capacity to manipulate and control the world around them.[14]

It appeared to make sense that political, social, and economic relationships would increase in complexity and sophistication over time. After 1859, it became possible to theorise that mind and society alike were subject to evolutionary forces. That year did not only mark the publication of Darwin's *Origin of Species* and Samuel Smiles' *Self Help*; it was also the year in which human prehistory became a scientific reality. In 1859, the geologist Joseph Prestwich and the archaeologist John Evans respectively made reports to the Royal Society and the Society of Antiquaries of London in which they confirmed a great discovery. The French scholar Boucher de Perthes had found flint tools that were contemporaneous with extinct mammal fossils in a French gravel pit. In other words, humanity was much older than had been assumed. Human existence in deep antiquity was confirmed as earlier discoveries of old bones were reinterpreted, and as more expeditions and excavations were undertaken in the latter half of the nineteenth century. Details of the discoveries of these digs were eagerly seized upon by both the public and by philosophers.

John Lubbock, a polymath politician, financier, and philanthropist, made the Victorian connection between ancient humans and colonised peoples explicit. In 1865, he published the influential book *Pre-Historic Times*, which swiftly became an archaeological classic and went through numerous editions over the following half-century. What is telling here is the book's full title: *Pre-Historic Times as Illustrated by Ancient Remains and the Manners and Customs of Modern Savages*. On the one hand, Lubbock argued that the function of prehistoric tools could be inferred from a consideration of how modern 'savages' used their own stone tools. On the other, the very fact that these 'modern' people were still using stone tools demonstrated that their socio-technical evolution had been halted at an earlier stage than Europeans'. Lubbock was not the only scholar to make this link explicit. The geologist William Boyd Dawkins, the first excavator of Wookey Hole on England's south coast, drew similar conclusions about the relationships of fossil and extant humanity, as did William Sollas in the book *Ancient Hunters and Their Modern Representatives* (1911).[15] These and many others were agreed: the evolution of some communities had simply come to a stop.

Contemporary accounts of the behaviour of 'primitive' humanity were woven together with folk tales, origin stories, and histories to create accounts of how human society became increasingly complex and sophisticated. The French scholar, Emile Durkheim, for example, pulled accounts of Jewish life from the Old Testament together with information about the present-day life of the Berber peoples of Algeria. Combining these two 'primitive' societies, one past and one present, allowed him to develop his analysis of 'segmentary' agricultural societies. At the time the Algerian information was gathered, France was involved in a war of conquest in North Africa; implicitly, Durkheim's story of development vindicated its efforts. Durkheim also used what he knew about the spiritual world of indigenous Australians and other First Peoples as the basis of his highly influential analysis of the role of religion in society. From his perspective, 'primitive religion' was 'better adapted than any other to lead to an understanding of the religious nature of man' because it was an expression of religion in its simplest form.[16] An absence of technological complexity was here being linked with what Europeans saw as an equally simple spiritual and social life.

European science, capitalism, and Christianity thus went hand in hand in a civilising mission that was supposed to kick-start the stalled evolution of the rest of the world's peoples. It would help them, to use a twentieth-century term, to 'develop'. Indigenous belief would be washed away, either through Christian conversion or by putting aside religion altogether. The process has been analysed by the postcolonial scholar Frantz Fanon. His discussion of the role of medicine, radio, and the veil in the Algerian wars of independence shows the extent to which science was deployed, both deliberately and unconsciously, in the attempt to compel indigenous Algerians to put aside traditional Islamic culture.[17] From the European side, this process consigned religion to the past. Only a child, or child-minded people (especially women), allowed their lives to be shaped by belief. Mature adults, who could put their understanding of the world to work by re-engineering it for the better, put their faith instead in science.

Unexpected turns

This chapter has emphasised the connection between science and Christianity in their shared 'civilising mission'. As such, it will perhaps have surprised some readers. After the initial Galilean skirmishes of the 1610s–1630s, the eighteenth and nineteenth centuries are usually considered to be the period in which Christianity gave way to science, with the story of Darwin being its most powerful and significant event. Our account, however, has more modestly pointed towards a gradual repositioning of science and faith within public and private spheres, respectively. This is not to say that science did not change during the period. It changed a great deal, becoming more expert and gaining an identity as a thing in its own right. No longer was there merely a variety of specific research activities; now there was 'science', practised by 'scientists'. In the course of these trends, identity conflicts did emerge. If a laboratory was the place to do science, then by definition, a vicar's study was not. If a scientific degree was the route to expertise, then by definition, a theological one was not.

The scientist T. H. Huxley was responsible for heating up what was in part a natural social process. We will explore the polemics of Huxley and his allies in the next chapter. From a humble background, Huxley was dismayed by the unholy alliance of class and church that gave cultural authority to bishops and their ilk. He was committed to the professionalisation of science as a meritocratic means by which men like himself could have their say, and he was only too glad to weigh in with the supposed superiority of science over religion in order to make his case.[18] He was surrounded by a group of like-minded men who dined and discussed regularly over a period of about thirty years: the 'X Club'.[19] Their philosophy was sometimes called scientific naturalism, a belief that everything in the natural world could be reduced to naturalistic explanations.

Not all the X Club's members thought the same, however. Lubbock saw science as a gift to reform the church and join with it in the joint enterprise of moral and intellectual education. In this, he supported the publication of an 1860 volume of essays that attempted to engage theology with the science of the day.[20] Contributors included Frederick Temple, who would later become Archbishop of Canterbury, as well as Professors of Greek and Geometry. The contents were rather uneven, but the book had a noteworthy impact, selling 22,000 copies in two years. This was most likely due to its timing, coming four months after Darwin's *Origin of Species* (which title it out-sold for two decades). It linked to Darwin's work by commending the understanding of the 'self-evolving power of nature' rather than an interventionist God. Lubbock, then, is an example of the complex relations between science and faith in Victorian Britain. As his biographer has suggested, Lubbock lived happily with an intellectual marriage of science and religion and suffered no crisis of faith or religious doubt.[21] Indeed, religion was a chapter in his book *The Pleasures of Life* (1887).

In part, the intention in taking this unexpected turn, towards connection rather than conflict, has been to avoid repetition. Over the past two generations, historians of science have examined lives like Lubbock's and public debates such as the book he supported to demonstrate the complex and nuanced theological responses to Darwin's theory and the broader ambitions of scientists. Most recently, scholars have begun to show that equally subtle responses occurred around the globe, in the context of Islam and other religions.[22] The further reading at the end of this chapter suggests some places to begin for readers unfamiliar with this historiography. Another reason for focusing on the consanguinity of science and faith in the civilising mission is that it makes sense of at least two of the case studies in Part II of this book. In the chapter on AI, we demonstrate how narratives of technological salvation (for the chosen few) emerge from this shared historical culture. In the case of Covid-19, too, we see how faith leaders and scientists came together to promote a 'rational' response to the pandemic—with some regrettable side effects.

There is yet one more unexpected turn to discuss in regard to the nineteenth century. Its ending saw multiple outbreaks of new religion and weird science on both sides of the Atlantic. Seventh Day Adventists, Jehovah's

Witnesses, Mormons, Christian Scientists, and Spiritualists all gave the lie to the idea that the lure of religion was fading. Meanwhile, respected scientists explored the psychical and paranormal potentialities of the latest physics.[23] X-rays, wireless waves, and other new phenomena suggested communications with realms beyond ordinary human perception. The Society for Psychical Research counted many noted scientists amongst its members.

The support that technology could provide to human spiritual growth was a key theme of this period and one that was closely connected with the possibilities of exploring space. The American geologist Louis Gratacap, for example, produced a scientific romance, *The Certainty of Future Life in Mars* (1903), which turned on the capacity of wireless telegraphy to contact other planets and to communicate with the once-human spirits now enjoying their afterlife on the red planet.[24] Chapter 4 will examine this theme in more detail. For now, it is intriguing to note that one of the very first photographs of the Moon was taken in March 1840 by John William Draper, the son of a Wesleyan clergyman. Forty years later, he would go on to write the book credited with popularising the conflict thesis, *The History of the Conflict Between Religion and Science*. As Chapter 3 will show, however, his work fits firmly within an older tradition of weaponising tensions, not between religion and science, but between contending denominations of faith.

Further reading

Geoffrey Cantor, *Michael Faraday: Sandemanian and Scientist: A Study of Science and Religion in the Nineteenth Century* (Basingstoke: Palgrave Macmillan, 2016)

Raewyn Connell, *Southern Theory: The Global Dynamics of Knowledge* (Cambridge: Polity Press, 2007)

Bernard Lightman and Sarah Qidwai, eds., *Evolutionary Theories and Religious Traditions: National, Transnational, and Global Perspectives, 1800–1920* (Pittsburgh: University of Pittsburgh Press, 2023)

Iwan Rhys Morus, *How the Victorians Took Us to the Moon* (London: Icon, 2022)

Sujit Sivasundaram, *Nature and the Godly Empire: Science and Evangelical Mission in the Pacific, 1795–1850* (Cambridge: Cambridge University Press, 2005)

Notes

1. 'Nikola Tesla Shows How Men of the Future May Become as Gods', *New York Herald*, 30 December 1900.
2. Ole Peter Grell and Roy Porter, eds., *Toleration in Enlightenment Europe* (Cambridge: Cambridge University Press, 2000).

3. James Hannan, *God's Philosophers: How the Medieval World Laid the Foundation for Modern Science* (London: Icon, 2009).
4. Jan Golinski, *The Experimental Self: Humphry Davy and the Making of a Man of Science* (Chicago: University of Chicago Press, 2016).
5. Sharon Ruston, 'When Respiring Gas Inspired Poetry', *The Lancet* 381 (2013): 366–367, https://www.thelancet.com/journals/lancet/article/PIIS0140-6736(13)60157-9/fulltext.
6. Thomas Dixon, Geoffrey Cantor and Stephen Pumfrey, eds., *Science and Religion: New Historical Perspectives* (Cambridge: Cambridge University Press, 2010).
7. Ruth Barton, '"Men of Science": Language, Identity and Professionalisation in the Mid-Victorian Scientific Community', *History of Science* 41 (2003): 73–119, https://doi.org/10.1177/007327530304100103.
8. Quoted in Iwan Rhys Morus, *How the Victorians Took Us to the Moon* (London: Icon, 2022).
9. Quoted in Iwan Rhys Morus, *How the Victorians Took Us to the Moon* (London: Icon, 2022); see also Matthew Stanley, 'By Design: James Clerk Maxwell and the Evangelical Unification of Science', *British Journal for the History of Science* 45.1 (2012): 57–73, https://doi.org/10.1017/S0007087410001548.
10. Samuel Smiles, *Self Help: With Illustrations of Character and Conduct* (London: Maxwell, 1859), p. 5.
11. Aileen Fyfe, *Science and Salvation: Evangelical Popular Science Publishing in Victorian Britain* (Chicago: University of Chicago Press, 2004).
12. Sujit Sivasundaram, *Nature and the Godly Empire: Science and Evangelical Mission in the Pacific, 1795–1850* (Cambridge: Cambridge University Press, 2005).
13. Auguste Comte, *The Positivist Philosophy of Auguste Comte* (Cambridge: Cambridge University Press, [1853] 2009).
14. William Whewell, 'On the General Bearing of the Great Exhibition', in *Lectures on the Results of the Great Exhibition*, Royal Society of Arts (London: David Bogue, 1852).
15. Matthew Goodrum, 'The Idea of Prehistory: The Natural Sciences, the Human Sciences and the Problem of Human Origins in Victorian Britain', *History and Philosophy of the Life Sciences* 34 (2013): 117–146, https://www.jstor.org/stable/43831770.
16. Emile Durkheim, *The Elementary Forms of Religious Life* (London: George Allen & Unwin, 1912), p. 13.
17. Frantz Fanon, *A Dying Colonialism* (New York: Grove, 1967).
18. Frank M. Turner, 'The Victorian Conflict between Science and Religion: A Professional Dimension', *Isis* 69 (1978): 356–376, https://www.journals.uchicago.edu/doi/abs/10.1086/352065.
19. Ruth Barton, *The X Club: Power and Authority in Victorian Science* (Chicago: University of Chicago Press, 2019).
20. John W. Parker, ed., *Essays and Reviews* (London: John W. Parker, 1860).
21. J. F. M. Clark, 'John Lubbock, Science, and the Liberal Intellectual', *Notes and Records of the Royal Society of London* 68.1 (2014): 65–87, https://doi.org/10.1098/rsnr.2013.0068.
22. Bernard Lightman and Sarah Qidwai, eds., *Evolutionary Theories and Religious Traditions: National, Transnational, and Global Perspectives, 1800–1920* (Pittsburgh: University of Pittsburgh Press, 2023).
23. Richard Noakes, *Physics and Psychics: The Occult and the Sciences in Modern Britain* (Cambridge: Cambridge University Press, 2020).
24. Iwan Rhys Morus, *How the Victorians Took Us to the Moon* (London: Icon, 2022).

3

Battling with history

The uses of history

If, then, faith and reason lived happily together in the West for most of recorded history, when did the 'war' between science and religion break out? In this chapter, we will argue that conflict did not commence until the late nineteenth century and that even then, it did so in a rather limited way. Active hostilities did not begin until the latter part of the twentieth century and did so in the broader context of some radical shifts in the role played by science in public life. The wide public, political, and institutional support for scientific and technological research that had characterised the first half of the twentieth century, and which was intensified by the role played by science during the world wars, was shaken by a number of post-war developments. Specific threats or controversies relating to new technologies (nuclear weapons and energy, genetic modification, food safety, nanotechnology, artificial intelligence) began to appear. They emerged within a broader context of growing concern regarding the impact that industrial society was having on the global environment.

Scientists' capacity to respond effectively to these crises was damaged by the perception that they were produced, at least in part, by declining levels of scientific literacy in the West. A pivotal report on the need to improve the public understanding of science, produced by the UK's Royal Society in 1985, is just one example of this concern.[1] But within the debates surrounding these and other issues, a notable development occurred. Themes that could be characterised as religious in nature (questions of morality, whether something 'should' be done, concerns about 'playing God') were often counterposed to the rationality, efficiency, or efficacy of science in solving the age-old problems (famine, pestilence, pollution, war) faced by humanity. These modern debates balanced morality against rationality, while at the same time, participants consistently placed themselves in an imagined history of an existential struggle between faith and knowledge that dated back

Science, Religion, and the Human Future. Amanda Rees et al., Oxford University Press. © Amanda Rees, Franziska E. Kohlt, Tom McLeish, Charlotte Sleigh, David Wilkinson (2025).
DOI: 10.1093/9780191995316.003.0004

at least to the early Christian Church, if not beyond. This is the political and cultural context within which the claim of a 'war' between science and religion was deployed.

The role of religion in the West was also in flux in this period. As Chapter 2 has shown, attitudes towards the role of faith in public life had shifted considerably by the turn of the twentieth century, especially for people that were considered to be part of the scholarly elite. While they did not necessarily eschew faith, members of these groups increasingly confined the expression of religious belief to their private lives. In particular, divine action was no longer considered to be an adequate or relevant explanatory factor when accounting for the natural world. Meanwhile, even as philosophers and social theorists began to categorise religion as an evolutionary 'stage' through which societies would pass as they progressed towards the development of a rationally organised civilisation, new forms of religious belief such as Christian Science and Spiritualism were coming into being. These drew explicitly on recent science and technology to support the existence and experience of spiritual life. Some parts of social life were becoming more secular, but this was unevenly experienced across different communities and different geographical and conceptual spaces. As many historians have shown, those 'boundaries'—when examined closely—are sometimes clearly drawn, but often extremely blurred, and occasionally wholly non-existent.[2]

It is still certainly the case that there are events in Western history that marked clear points of tension, where natural philosophers, scientists, and other public figures recognised themselves to be in conflict with religious ideas, institutions, and individuals. The most obvious example here, and the one that—along with the 'Galileo affair'—is most often cited when discussing these questions, is the public deliberation and disputation that followed the publication of Charles Darwin's *Origin of Species* in 1859. But, as this chapter will argue, the events that occurred a century or so later are just as important, if not more so, to the understanding of twenty-first-century concerns about the relationship between science and religion. Particularly important, as we will see, is the shifting post-war public perception of the role of science in society, including the growth of the counterculture in the 1960s, the environmental concerns of the 1970s, and the cutbacks in funding to higher education of the 1980s, as well as the 'Science Wars' of the 1990s. What is especially fascinating, however, is the way that the *history* of science was operationalised by participants in these discussions as a means of demonstrating the antiquity of science/religion animosity, in a way that often showed very little regard for the historical context of the events in question.

As a result, this chapter is in many ways less about *history* than it is about *historiography*: that is, how history is written, perceived, and used in public narratives and debates. In this sense, it is appropriate for us to begin with one of the first 'official' histories of science—Thomas Sprat's *History of the Royal Society* (1667)—commissioned by the Society out of concern for its public reputation. In his account of the Society's principles and origins, Sprat explored the political and practical accommodation that England's Royal Society had made with Anglicanism and the monarchy in the late seventeenth century, in the aftermath of the country's civil war.

Restoration, reformation, and the Royal Society

The Royal Society was originally founded in November 1660; its patron was King Charles II, who had reclaimed the English crown in April of that year. While the restoration of the monarch might officially have ended the struggles over the governance of the country that had begun when Royalists and Parliamentarians went to war in 1642, political, cultural, and religious tensions understandably remained high. The nascent Society, despite its royal patronage, found itself in a rather uncertain position. It had no financial stability, being dependent on subscriptions from its members, many of whom took no, or very little, part in its business. It was potentially in danger of being tainted by association with political elements (materialism, radicalism, liberalism) that were regarded with great suspicion in the post-revolutionary period. One way of heading off this potential threat was to make an explicit and public statement as to the Society's aims, activities, and remit. Thomas Sprat, fellow of Oxford's Wadham College and prebendary of Lincoln Cathedral, was commissioned to write such a statement, a history of the Society, under the supervision of one of its founding members, John Wilkes.[3]

The history was an opportunity for the Society to position itself in this new world. What relationship did it believe itself to have with the restored monarchy and the reformed Church? How was it to avoid accusations of scepticism (associated with atheism and immorality) or enthusiasm (linked with dogmatism) in finding a secure place for itself in public life? The solution put forward by Sprat and Wilkes was to focus on the *methodology* adopted and avowed by the Society, and to show how that methodology both reflected and supported the monarchy and the Church. What they were trying to do, in effect, was to show that the empirical method was in fact a means of putting Anglicanism into practice. Just as Anglicanism had reformed the Church, so would the Society reform natural philosophy.

Sprat was putting forward a theological argument as much as writing a history of the first learned society. Chapter 1 of this book explored Robert Grosseteste's argument that through reasoning would come redemption. Sprat's suggestion was even more specific—that it was through the application of the empirical method that redemption might be achieved. That is to say, through the fall of man, humanity had lost access to the true understanding of the world originally granted by God to Adam and Eve. Neither reason nor human senses were sufficient foundations on which to rebuild this lost knowledge: both depended on human frailty and were therefore unreliable. Through experiment and the empirical method, however, humanity could overcome these bodily limitations. New scientific instruments acted as means through which to extend and support human senses. The information derived from these investigations was then to be interrogated and interpreted: the 'truth' or otherwise of knowledge claims had to be tested by means of collective debate. Only in this way could the fallibility of the individual intellect be overcome and the prelapsarian insight into the workings of the world be regained.

In this way, Sprat tied the practice of the experimental method to the principles of the reformed Church. On this reading, Henry VIII's break with Rome was inspired not by the desire for a son, but by the need to liberate Christians from the tyranny of the Pope. True knowledge of either the spiritual or natural world could not be derived from one man's dictat but had to emerge from reasoned debate—debates undertaken not for debate's sake (as in the case of the scholastics) but drawing upon and testing observable evidence. Just as importantly, and despite the Clarendon Code (which restricted the access of non-Anglicans to public life), the Royal Society could be ecumenical, drawing in members from many different walks of life and backgrounds. As Sprat put it, within its meetings:

> men of disagreeing parties, and ways of life, have forgotten to hate, and have met in the unanimous advancement of the same *Works*. There the *Soldier*, the *Tradesman*, the *Merchant*, the *Scholar*, the *Gentleman*, the *Courtier*, the *Divine*, the *Presbyterian*, the *Papist*, the *Independent* and those of *Orthodox Judgment*, have laid aside their names of distinction.[4]

The Royal Society thus consciously positioned itself as an ally of the Anglican Church, while strongly implying that its experimental method could represent a path through which dissenters and Catholics might find their way to reasoned enlightenment. Going further still, empirical evidence could be adduced to demonstrate the divinity of Christ. What were miracles, after all, but an empirical demonstration of divine nature? As historian Peter Wood put it in his commentary on the *History*, 'Sprat characterised Christ as the ideal experimental philosopher, practicing the same method as the Fellows of the Society, but with infinitely greater ability'.[5] After all, it was only through a thorough knowledge of the laws of nature that genuine miracles could be either conceived or recognised.

It is important to remember that not all Sprat's contemporaries would have agreed with his identification of the experiential world as a path to God. And it was certainly not the case that all members of the Society would have recognised, much less agreed, with the account that Sprat gave of the relationship between the Society and the Anglican Church of England. As many historians have shown, Sprat's account of the origins and aims of the Royal Society was often oversimplified, highly selective, and frequently partial.[6] But the aim of his *History* was to carve out a more secure cultural and political (and eventually financial) place for the new philosophy within English life. As such, the *History* represented an important step in the institutionalisation of science within public life, an achievement which at that time depended on the ability of practitioners to demonstrate the intimacy of the relationship between science and a particular sect of Christianity. This question of sectarianism was to become more important in the science/religion debates that flourished in the aftermath of the publication of Darwin's *Origin of Species*.

Draper, Tyndall, and Darwin

In July 1858, a collection of Charles Darwin's notes and an essay by Alfred Russel Wallace were read to the Linnean Society in London. Together, these papers established the basis of the theory of evolution through natural selection, at which the two men had arrived independently. A far more detailed account of the theory, Darwin's *Origin of Species*, was published in November of the following year and swiftly became a best-seller in Victorian Britain. Evolution itself was not, of course, a new concept—but the theory of natural selection, particularly as applied to humanity, was both novel and compelling. In essence, the theory relied on three factors: first, that more organisms are born than can survive to reproductive age; second, that each organism will inherit from its parents a subtly different range of characteristics; and third, that those organisms which survive will be those whose inherited characteristics are best fitted to the vagaries of the environment in which they find themselves (better camouflage, a beak that can open a wider range of seeds, a longer neck). As a result, over time and as environments change, or as animals spread to unfamiliar surroundings, new species will emerge that reflect these advantages.

In a number of ways, this theory seemed to conflict with many Christians' understanding of God's relationship with the world. In the first place, the fact that natural selection operated as an ostensibly random process, being dependent on unpredictable changes in the environment and on the appearance of natural variation, seemed to contradict the idea that God had designed the world with a place for everything. In the second, the fact that so many individuals were born simply to die seemed wasteful, a Darwinian world of ruthless and bloody competition apparently incompatible with the idea of a loving and benevolent deity. Finally, of course, the fact that natural selection treated humanity as just one more species amongst many, and that our own origins seemed likely to lie closer to the apes than the angels, conflicted directly with the scriptural account of humanity's special creation. These three areas of tension were difficult to reconcile with natural theology and the argument from design, which underpinned many of the arguments that had historically been used to motivate and inform theological engagement with the natural world. But while Darwin himself was concerned about the potential impact natural selection might have on religious belief, in particular fearing that it might be taken as a support for atheism, the actual impact his theory had on the faithful was rather more nuanced than many accounts, past and present, imply.[7]

The years that followed the publication of the *Origin* saw a number of events that have gained legendary status in the narrative of the 'war' between science and religion. Most significant in terms of involving apparent face-to-face conflict was probably the encounter between 'Darwin's Bulldog' T. H. Huxley and 'Soapy Sam' Wilberforce, the Bishop of Oxford. The nicknames are themselves a telling and significant part of the historiography in the way that they contrast the brave tenacity of the bulldog with the slippery unpredictability of soap. In May 1860 (the same year that Andrew Dickson White helped found Cornell University in the USA as 'an asylum for Science— where truth shall be sought for truth's sake, not stretched or cut exactly to fit Revealed Religion'), the annual meeting of the British Association for the Advancement of Science was held in Oxford.[8] At this meeting, an American scholar, John William Draper, whom we last saw photographing the Moon in Chapter 2, read a paper on the evolution of the European intellect. In the discussion that followed his paper, Huxley and Wilberforce exchanged some forceful views on the subject of Darwinian evolution. No account of what was said was recorded at the time, although there are a number of retrospective versions in letters, diaries, and newspaper accounts. There is considerable variation amongst these sources. Not only do they offer a range of different versions of what actually happened, but they also reach different conclusions on who 'won' the debate.

Nevertheless, in terms of the science/religion debates, the story is treated as relatively simple. Rather than engaging with any matters of fact, Wilberforce is said to have relied on oratorical tricks to discredit natural selection, sarcastically asking Huxley if he would rather be descended from the apes on his grandmother's or his grandfather's side of the family. Huxley's much-mythologized response was to declare that he would far rather have a monkey for an ancestor than be related to a man who used rhetoric to hide from scientific truth. The success of this dramatic 'Great Debate' clearly reflects the broader structure of the narratives used to support the theory of the 'science/religion' conflict: hostility and unwillingness on the part of 'religion' to face up to any facts placed in front of it by 'science' that might conflict with doctrine.[9] Faith is conflated with unthinking acceptance of received wisdom, science with rational questioning, and independent thought.

John Draper went on to develop the paper that had preceded the debate into a book, *A History of the Intellectual Development of Europe*, which was published in 1862. In turn, the book had a significant impact on the next events in the emergence of the 'conflict thesis'. It was extensively used by the Irish physicist John Tyndall in his presidential address to the British Association's Belfast meeting in 1874, which further entrenched the developing narrative of the 'eternal' conflict between science and religion. New presidents of the British Association were appointed annually and by tradition, the new incumbent's address consisted of an overview and evaluation of the key scientific advances of that year. Tyndall, however, took a different tack. Rather than focusing on the immediate past, he confronted his audience with his own interpretation of the whole history of science, from the ancient philosophers to the present day.

The story Tyndall told was one in which the proper understanding of the material world and the basic building blocks of life itself were consistently obfuscated by the Christian Church. Classical philosophers such as Leucippus and Democritus, Tyndall suggested, had done a competent job of thinking about the fundamental components of living matter, but this train of thought had been thoroughly derailed in Europe, if not elsewhere in the world (as Draper's book had argued), by the appearance of Christianity. From Tyndall's perspective, it was not until after Copernicus' heliocentric demonstrations, followed by the martyrdom of Giordano Bruno and Galileo's forced recantation at the hands of the Inquisition as well as Newton's careful elaboration of the mathematics underpinning gravity, that Darwin came to present the latest and most telling refutation of the clerics.[10] From this point on, Tyndall concluded, all claims that pretend to objective knowledge must submit to the control of science: religion must not attempt to 'intrude on the region of *knowledge*, over which it holds no command', but must instead confine itself to the realm of emotion 'which is its proper sphere'.[11]

The notion that all real questions about knowledge were the proper domain of science alone came to be known as the principle of 'scientific naturalism'. Tyndall's claim for the supremacy of science was based in its methods. Experiments in laboratories enabled scientists to produce and identify accurate knowledge about the natural world. The conduct of scientists themselves was another essential ingredient; their ability to discipline their own bodies and minds in the course of proper experimentation was the essential precondition to exerting control over nature. There was thus a moral element to Tyndall's secularism. His history of science was based on the biographies of great men. These were defined as great not because of their genius, but because of their virtue, exhibited through their capacity for self-discipline and self-sacrifice in the name of humanity. As a result, Tyndall saw virtue not as innate, but as something that could be cultivated, meaning that—at least potentially—anyone was capable of participating in the drive to extend objective, experimental knowledge of nature. Understandably, such a strong claim to a singular path to truth infuriated and disturbed many of those who heard it: not just clerics like Bishop Wilberforce and John Henry Newman, but also physicists such as Oliver Lodge, who noted the 'dogma' that drove Tyndall's speech. The physicist James Clerk Maxwell satirised Tyndall's version of science history in verse:[12]

> In the very beginnings of science, the parsons, who managed things then,

Being handy with hammer and chisel, made gods in the
likeness of men;
Till commerce arose, and at length some men of exceptional
power
Supplanted both demons and gods by the atoms, which last
to this hour.[13]

The year 1874 also saw the publication of John Draper's *History of the Con-flict between Religion and Science*. This was the most explicit account to date of the 'conflict' thesis, and the source of its name. It treated the history of science as a 'narrative of the conflict of two contending powers, the expansive force of the human intellect on the one side, and the compression arising from traditional faith and human interests on the other'.[14] As Tyndall had done, Draper identified the origins of science in classical philosophy and the repression of science in the emergence of Christianity. His book was followed in 1896 by Andrew Dickson White's *A History of the Warfare of Science with Theology in Christendom*, which—while acknowledging that Christianity had in the past supported scientific and philosophical engagement—stressed that any interference by religion in scientific investigation would inevitably lead to disaster. The men had been encouraged to write the books by the publishing brothers Edward and William Youmans, whose sales justified the brothers' belief in public appetite for the subject matter. The popularity and influence of the books were so great that historians and other commentators have credited them with originating the 'conflict thesis' itself.[15]

It is certainly the case that Draper's and White's stories have continued to resonate in modern public discussions regarding the relationship between religion and science, despite consistent and regular efforts by historians of science to correct some of their mistaken assertions. Not all were based in historical fact. Episodes such as the Inquisition's treatment of Galileo, the Church's attitude to the examination of dead bodies, the medieval belief in a flat Earth, and the debates about the age of the Earth were creatively retold. So too was the dispute between Huxley and Wilberforce at Oxford. There is a great irony when it comes to considering the impact that the work of Draper and White had on the science/religion debate; while both men misrepresented history, they have also themselves been misrepresented by history.

The 'threat' that both writers identified came not from 'religion' or 'Christianity' per se but from specific sects within the Christian tradition.

Their attacks were aimed not at the followers of a 'rational' Protestant religion, but at Catholicism and the dogmatic dissenting sects which, on their reading, equated faith with unquestioning belief. Like Sprat in the late seventeenth century, these nineteenth-century writers saw no conflict between science and the rational Protestant religion. Even other religions were preferable to Catholicism; both Draper and Tyndall singled out Islam and Buddhism as congenial to the ideas and principles of modern science. If anything, they believed that science could help clarify theological understanding and would, in fact, potentially defend religious belief by helping to place it on firmer conceptual foundations.[16]

History of science in the twentieth century

Forty years after White's book appeared, an American graduate student, Robert Merton, submitted his doctoral thesis. This thesis—appearing two years later in print as *Science, Technology and Society in Seventeenth-Century England*—represented a significant reworking of an idea originating with the sociologist Max Weber. Weber had argued that Protestantism had played a significant role in the emergence of capitalism, shaping attitudes to work, investment, and self-discipline. In turn, Merton correlated Protestants' (and particularly Puritans') ethics with the ethos of the emergent experimental sciences, arguing that the religious backgrounds and values of these people had similarly encouraged and inspired their engagement with the natural world. Merton did not claim that Protestantism *caused* the 'scientific revolution', nor even that the existence of Puritanism was an essential precondition to the emergence of the new philosophy. Instead, he argued that 'the cultural soil of seventeenth century England was peculiarly fertile for the growth and spread of science' and that a significant aspect of that fertility was found in the Puritan ethos, which in the present day still resonates with idealised scientific virtues, values, and norms.[17] Merton's thesis has been the focus of considerable attention and debate since its publication. The fact that it became and remains the subject of serious discussion is a clear indication that, in scholarly communities during the early twentieth century, there was no consensus as to the existence of inevitable conflict between religion and science.[18]

At no point in the twentieth century did professional historians of science find empirical support for the narratives produced by Tyndall, Draper, and others, with their origins in Classical Greece, their period of ecclesial suppression, and their eventual renaissance in the Scientific Revolution. Consider, for example, the tremendous scholarship on the history of cosmology by French philosopher and physicist Pierre Duhem. Composed in the early twentieth century, this established the fertility and dynamism of medieval natural philosophy as well as the extent of Church support for these activities.[19] Similarly, Alfred North Whitehead's Lowell lectures of 1925 acknowledged that the rise of modern science had disrupted Western Christianity, yet also insisted that any attempt to characterise the relationship as a 'conflict' represented a serious oversimplification.[20] In the post-war period, historians followed on from Merton's characterisation of Puritan-friendly science; Richard Westfall, for example, showed how Protestant theology influenced the emergent new philosophy.[21] During the 1960s and 1970s, considerable scholarship was devoted to examining the nature of the evolution debates, culminating in James Moore's influential *Post-Darwinian Controversies* (1979). This, in turn, inspired a number of subsequent studies charting the complexity of the religious response to Darwin's theory of natural selection on both sides of the Atlantic.[22]

While twentieth-century historians examined Darwin's impact on nineteenth-century theologians, twentieth-century evolutionists continued to engage directly with theology. In 1947, for example, Alister Hardy, Linacre Professor of Zoology at Oxford, set up a new Department of Zoological Field Studies, bringing together the Bureau of Animal Population (led by Charles Elton) and the Edward Grey Institute of Ornithology (under David Lack), as well as creating space for a new appointment (and future Nobel laureate), Nikolaas Tinbergen. The work and achievements of these four scholars were fundamental to the development of evolutionary ecology in the twentieth and twenty-first centuries. David Lack in particular took forward Darwin's interest in the finches of the Galapagos Islands through an investigation of beak structure in the context of diet and environment. It was, in fact, Lack's work rather than Darwin's which ensured that 'Darwin's finches' became the textbook exemplar of how natural selection worked in terms of adaptive radiation.

Lack saw no conflict between his devout Christian faith and his study of evolution. In 1957, he produced a book, *Evolutionary Theory and Christian Belief*, in which he provided answers to the triple threat that the theory of natural selection had seemed to pose for natural theology. He argued that natural selection did not actually operate randomly, as well as dealing with the apparent brutality of the natural world and even addressing the question of miracles. His colleague and head of department, Hardy, went further, arguing that accounts of evolutionary behaviour had to take consciousness and agency into account, and that religious belief, or the ability to become aware of transcendence, had been selected for in human evolution. Most significantly, Hardy made the case that the scientific study of the 'biology of God' was not only a feasible endeavour, but also an essential task for scientists to address if they wanted to understand the nature of humanity.[23]

Many scientists have seen no conflict between scientific practice and religious belief, and almost no professional historians of science have considered the 'war' between science and religion to have any basis in empirical evidence. Yet, the conflict myth has been able to exert a powerful traction over public debates on the place of science and religion in public life in the late twentieth and early twenty-first century. Why?

Twentieth-century science in public culture

The cause of the conflict myth's traction can be found in the changing status and position of science in public culture in the West during the twentieth century. Technoscientific innovation had fundamentally changed human life in the West by the end of the twentieth century, largely for the better: electrifying homes, industrialising agriculture, exploring space, and managing human reproduction. Biomedicine and better understandings of public health meant that people were living longer and living better; improvements in communication technology made commerce and kinship easier to build and manage; and chemical and biological innovation was increasing the resilience and reliability of the food supply. But there was another side to these positive developments. As early as the 1930s, the science fiction writer H. G. Wells had been sufficiently concerned about the increasing speed and scope of technological innovation to call for the inauguration of 'Professors of Foresight', whose job it would be to predict the impact of science on society. He noted, in a talk he gave to the BBC's National Programme in 1932, that it was not enough to invent the motor car if you couldn't also predict the traffic jam or the 'motor bandit'.[24] The unintended consequences of innovation were starting to be felt, and to disrupt, the positive story of scientific progress.

By the early 1960s, these consequences of scientific and technological innovation were becoming clearer. In 1958, for example, the St Louis Citizens' Community for Nuclear Information had initiated the 'Baby Tooth

Survey' in order to find out the extent to which atomic testing was affecting the bodies of children born in the United States, inspiring a number of other similar surveys globally. The first Aldermaston March in protest against nuclear weapons took place in the United Kingdom in 1958, becoming an annual event in 1959. It was followed internationally by other protests. In the US, Rachel Carson published *Silent Spring* (1962), outlining the devastating ecological impact of the uncontrolled use of the pesticides that supported the agricultural revolution. In the UK, the book's publication coincided with the 'Smarden incident', an extremely serious toxic waste spill in Kent which became emblematic of the unforeseen consequences of technological development.[25] The 1980s saw a number of disasters involving huge loss of human life and environmental damage: the protracted Ethiopian famine and the Chernobyl inferno; the explosion of the space shuttle *Challenger* and the Bhopal chemical disaster; and oil spills in the Niger delta, the Amazon watersheds, and the Alaskan coast. These were just a few of the technological and natural calamities that marked the decade. Far from ensuring food security, technoscientific innovation seemed to be rendering the food supply far riskier—Spain's toxic oil syndrome, Alar in US apples, BSE in British beef. Partly, these events were due to the complexity of global supply chains; they were also reflective of issues relating to the scientific industrialisation of food production. Faith in the capacity of science and technology to securely manage humanity's relationships with the natural world was severely damaged by the fallout from these and other events.

The 1970s and 1980s saw challenges to the narrative of control over nature that had underpinned the stories of scientific progress, and a new emphasis on the 'limits to growth'—the finite nature of natural resources. The 'Earthrise' picture, taken on Christmas Eve 1968 by the Apollo 8 astronauts during their circumnavigation of the Moon, did a great deal to help the public conceptualise the fragility and beauty of life on Earth. The following year, Buckminster Fuller's book *Operating Manual for Spaceship Earth* emphasised the need for self-restraint on the part of industrialised nations. He cautioned his readers that:

> we can make all of humanity successful through science's world-engulfing industrial evolution, provided that we are not so foolish as to continue to exhaust in a split second of astronomical history the orderly energy savings of billions of years' energy conservation aboard our Spaceship Earth. These energy savings have been put into our Spaceship's life-regeneration-guaranteeing bank account for use only in self-starter functions.[26]

Alongside the environmental movement, and linked through publications such as the 'Whole Earth Catalog' inspired by Stewart Brand, was the broader movement that became known as the 'counter-culture': groups inspired by the movements for civil, equal, and sexual rights, opposition to war in general and the war in Vietnam in particular, experimental use of psychoactive drugs, the rise of alternative lifestyles, and the rejection of traditional sources of authority. According to theorist Theodore Roszak, at the heart of this movement was the desire to subvert the scientific worldview itself, treating it as a reductionist approach which needed to be replaced with a holistic, spiritual stance if humanity was to survive a coming Apocalypse.[27] At the same time, a surprising number of the US and UK populations were claiming to have seen flying saucers and even to have encountered aliens in person.[28] Others were pursuing parapsychological experiments, owning pet rocks, watching Uri Geller bend spoons on live TV, or avidly reading Immanuel Velikovsky's claims to have found scientific proof of the miracles recounted in Genesis and Exodus. There is a fascinating combination here as people (including scientists) embraced the idea of science in their efforts to produce experimental proof of their ideas while, at the same time, rejecting the rationalism of mainstream scientific thought.[29] For conventional scientists, these movements, when combined with the cuts made by US and UK governments to the public funding of higher education and scientific research in the 1980s, suggested that the place of science in public life was under threat.

Some scientists also felt threatened by what they identified as a growth in public hostility towards them. In some cases, this translated into direct personal threat from animal rights activists. In other cases, the threat was more inchoate: surveys of scientific literacy undertaken on both sides of the Atlantic showed that the populations of the US and UK were apparently ignorant about science. Many people could not explain what the scientific method was, or how to test a hypothesis. Perhaps bamboozled by a badly framed question, many were happy to agree with the statement that the sun revolved around the Earth.

Even worse, there appeared to be a growing threat to empirical rationality from within the academy. Scientists found themselves subject to scrutiny by sociologists, historians, and anthropologists. Thomas Kuhn had, back in 1962, published a book called *The Structure of Scientific Revolutions*, which argued that the history of science was not, as had been previously assumed, the process of steadily piling up more and more knowledge about the natural world, but instead a series of paradigm-shifting revolutions that kept changing the framework of what counted as knowledge. This insight was developed into what became known as the 'strong programme' in the sociology of science, where sociologists and historians began to examine the social nature of truth. This meant looking at how different scientific communities adjudicated the claims to truth. For many scientists, this felt uncomfortably close to suggesting that all truth was relative.

By this point, evolution had again become a matter for public debate. In 1949, John Paul Scott had coined the term 'sociobiology' to describe a new perspective on animal behaviour, one that looked at the biological roots of social behaviour. The entomologist E. O. Wilson popularised the term in 1975 with the publication of the book *Sociobiology: The New Synthesis*, inciting a controversy that was simultaneously fought out within the academic community (with the historians, sociologists, and anthropologists who were unhappy with the idea that their disciplines should, or could, now be subsumed within biology) and in public (with citizens who were affronted by the claim that social hierarchies of race, gender, and class could have a biological basis). Discussions of 'creationism' or 'creation science', together with more explicit religious opposition to the teaching of Darwin's theory of evolution through natural selection, had rumbled on in the United States throughout the twentieth century. By the 1990s, it had been overtaken by the proponents of 'intelligent design' (ID) who wanted their approach to be recognised as a valid scientific hypothesis. In 2004, a court case was brought against a Pennsylvania school board to prevent them from presenting ID as an alternative to evolution: one of those who testified in support of the scientific status of ID was the sociologist of science, Steve Fuller. For many scientists, the sociobiology debates had merged into what became known as the 'Science Wars', in which the continued survival of science was potentially at stake.

This was the intellectual and political context within which the 'new atheism' of the early twenty-first century emerged, discussed in Chapter 10. Their work presented science as having been threatened throughout its history by the forces of irrationality, sometimes specifically embodied in the Christian Church, sometimes more generally characterised as 'religion'. Like their nineteenth-century predecessors, they both framed and illustrated their philosophical arguments through the history of science, though lacking Draper's and Tyndall's honourable mentions for Islam and Buddhism. The hard line drawn between science and religion in the Anglophone West, though intended to support science's position in public life, has had dangerous consequences for both scientific research and public culture, as Part II will now show.

Further reading

Subhadra Das, *Uncivilised: A Science Historian Explores Ten Founding Ideas of Western Civilisation and Unearths Their Flaws* (London: Coronet, 2024)

Peter Harrison, *The Bible, Protestantism and the Rise of Natural Science* (Cambridge: Cambridge University Press, 1998)

Bernard Lightman, ed., *Rethinking History, Science, and Religion: An Exploration of Conflict and the Complexity Principle* (Pittsburgh: University of Pittsburgh Press, 2019)

Notes

1. *The Public Understanding of Science: Report of the Royal Society's* ad hoc Group (The Royal Society: London, 1985).
2. John Hedley Brooke, *Science and Religion: Some Historical* Perspectives (Cambridge: Cambridge University Press, 1991); Bernard Lightman, ed., *Rethinking History, Science, and Religion: An Exploration of Conflict and the Complexity Principle* (Pittsburgh: University of Pittsburgh Press, 2019); Thomas Dixon, Geoffrey Cantor, and Stephen Pumfrey, eds., *Science and Religion: New Historical Perspectives* (Cambridge: Cambridge University Press, 2010).
3. John Morgan, 'Religious Conventions and Science in the Early Restoration: Reformation and "Israel" in Thomas Sprat's *History of the Royal Society* (1667)', *British Journal for the History of Science*, 42.3 (2009): 321–344, doi:10.1017/S0007087409002179.
4. Thomas Sprat, *The History of the Royal Society of London for the Improving of Natural Knowledge* (London: Royal Society, 1667), p. 427, https://archive.org/details/b3032760x.
5. P. J. Wood, 'Methodology and Apologetics: Thomas Sprat's "History of the Royal Society"', *British Journal for the History of Science* 13.1 (1980): 1–26, p. 15, doi:10.1017/S0007087400017453.
6. John Morgan, 'Religious Conventions and Science in the Early Restoration: Reformation and "Israel" in Thomas Sprat's *History of the Royal Society* (1667)', *British Journal for the History of Science*, 42.3 (2009): 321–344, doi:10.1017/S0007087409002179.
7. David Livingstone, *Darwin's Forgotten Defenders* (Edinburgh: Scottish Academic Press, 1987).
8. David C. Lindberg and Ronald L. Numbers, eds., *God and Nature: Historical Essays on the Encounter between Christianity and Science* (Berkeley: University of California Press, 1986), pp. 2–3.
9. A later addition to the legend was Huxley's remark on hearing that Wilberforce had died from head injuries incurred by falling from his horse: 'For once, reality and his brains came into contact and the result was fatal'. See William Ashworth (2020) 'Scientist of the Day: Samuel Wilberforce'), Linda Hall Library, 7 September 2021, https://www.lindahall.org/about/news/scientist-of-the-day/samuel-wilberforce/, accessed 13 February 2025.
10. Ian Hesketh, 'The making of John Tyndall's Darwinian Revolution', *Annals of Science* 77.4 (2020): 524–548, https://doi.org/10.1080/00033790.2020.1808243.
11. John Tyndall 'Inaugural Address of Prof. John Tyndall, D.C.L, LL.D, F.R.S., President', originally published in *Nature* 10 (1874): 309–319, p. 318. Available at https://www.victorianweb.org/science/science_texts/belfast.html, accessed 13 February 2025.
12. Oliver Lodge, *Man and the Universe* (London: Doran, 1908), pp. 6–8; Richard T. Glazebrook, *James Clerk Maxwell and Modern Physics* (London: Cassell, 1896), ch. 6.

13. David Tyler, 'An Insight into Maxwell's Mind?', *Nature*, 472.38 (2011), https://doi.org/10.1038/472038b.
14. John William Draper, *History of the Conflict between Religion and Science* (London: Henry S. King, 1875), 'Preface'.
15. James R. Moore, *The Post-Darwinian Controversies: A Study of the Protestant Struggle to Come to Terms with Darwin in Great Britain and America, 1870–1900* (Cambridge: Cambridge University Press, 1981), pp. 19–49.
16. James Ungureanu, *Science, Religion and the Protestant Tradition* (Pittsburgh: University of Pittsburgh Press, 2019).
17. Robert Merton, 'Science, Technology and Society in Seventeenth Century England', *Osiris*, 4, pp. 360–632, p. 597, https://www.journals.uchicago.edu/doi/epdf/10.1086/368484.
18. For a discussion of the historiography of the Merton thesis, see Steven Shapin, 'Understanding the Merton Thesis', *Isis*, 79.4 (1988): 594–605, https://www.journals.uchicago.edu/doi/abs/10.1086/354487. Peter Harrison developed a nuanced elaboration of the connection between Protestant ethics and the emergence of modern science in *The Bible, Protestantism and the Rise of Natural Science* (Cambridge: Cambridge University Press, 1998) where he showed how changes in the way that the Bible was interpreted and read by Protestants helped encourage and motivate scientific inquiry.
19. Pierre Duhem, *Medieval Cosmology: Theories of Infinity, Place, Time, Void and the Plurality of Worlds* (Chicago: University of Chicago Press, 1985).
20. Alfred North Whitehead, *Science and the Modern World* (New York: Norton, 1925). Given the place of the Galileo narrative in the history of the conflict legend, it is interesting to note that on Whitehead's account, the worst that happened to Galileo was that he 'suffered an honourable detention and a mild reproof' from the Church (p. 2).
21. Robert Westfall, *Science and Religion in Seventeenth Century England* (London: Yale University Press, 1970/1958).
22. James R. Moore, *The Post-Darwinian Controversies: A Study of the Protestant Struggle to Come to Terms with Darwin in Great Britain and America, 1870–1900* (Cambridge: Cambridge University Press, 1981); Stephen C. Barton and David Wilkinson, *Reading Genesis after Darwin* (Oxford: Oxford University Press, 2009).
23. David L. Lack, *Evolutionary Theory and Christian Belief: The Unresolved Conflict* (London: Methuen, 1957); Alister Hardy, *The Divine Flame: An Essay towards a Natural History of Religion* (London: Collins, 1966); Alister Hardy, *The Biology of God: A Scientist's Study of Man the Religious Animal* (London: Cape, 1975); William Homan Thorpe, *Science, Man and Morals* (London: Routledge, 1965/2020).
24. BBC, 'H. G. Wells on Consequences of Technology' (1932), https://www.bbc.co.uk/videos/c1ejjg4w7x0o, accessed 13 February 2025.
25. J. F. M. Clarke, 'Pesticides, Pollution and the UK's Silent Spring: Poison in the Garden of England', *Notes and Records: The Royal Society Journal of the History of Science* 71(3): 297–327, https://doi.org/10.1098/rsnr.2016.0040.
26. Richard Buckminster Fuller, *Operating Manual for Spaceship Earth* (Baden: Lars Muller, 2008), p. 135.
27. Theodore Roszak, *The Making of a Counter Culture* (New York: Doubleday, 1969). For a critique of this approach, see David Kaiser and W. Patrick McCray, eds., *Groovy Science: Knowledge, Innovation and American Counter-culture* (Chicago: University of Chicago Press, 2019).
28. Greg Eghigian, *After the Flying Saucers Came: A Global History of the UFO Phenomenon* (Oxford: Oxford University Press, 2024).
29. Harry Collins and Trevor Pinch, *Frames of Meaning: The Social Construction of Extraordinary Science* (London: Routledge, 1982).

PART II
PRESENTS

4
Space

Wars, real and imagined

In February 1947, the same year that Kenneth Arnold reportedly saw 'flying saucers' circling Mount Rainier in the US state of Washington, Earth-evolved life left the planet for the first time. The species in question was the fruit fly, launched on a V2 rocket from the United States. A decade later, in 1957, the USSR sent the world's first artificial satellite into orbit around the Earth using inter-continental ballistic missile technology. This demonstration of Soviet technological virtuosity caused consternation in the United States and other western European nations. Not only was the satellite itself considerably bigger than anything the US had been planning to send up, but the USSR had made it plain that the American mainland lay within the reach of its weapons. Comfortable assumptions about Western scientific and technological superiority, already challenged by the successful test of a Soviet atomic bomb in 1949, had been abruptly shattered. By the time Yuri Gagarin became the first human to successfully orbit the Earth in 1961, the 'race to space' had become a key front-line in the Cold War conflict.

Direct human exploration of space was always closely linked to actual, as well as metaphorical, warfare. The Saturn V rockets in which American astronauts lifted off from the planetary surface were constructed under the direction of the engineer Werner von Braun. A civilian specialist working for the German military, he had been responsible for the development of liquid-fuel V2 rockets, built using slave labour from concentration camps and deployed by the Nazis in World War II. In the closing months of that war, as their troops flooded into Germany, both the US and the USSR tried to locate as many V2-experienced staff as possible with an eye to using them in future conflict. Von Braun and his immediate colleagues decided to surrender to the advancing American army in preference to capture by the Soviets. Relocated to the US, he continued to work on the V2 rockets for military purposes while planning for their eventual use to 'conquer' space.[1]

Science, Religion, and the Human Future. Amanda Rees et al., Oxford University Press. © Amanda Rees, Franziska E. Kohlt, Tom McLeish, Charlotte Sleigh, David Wilkinson (2025).
DOI: 10.1093/9780191995316.003.0005

While literal human exploration of space began in that post-war context, the *idea* of outer space and the concept of extra-terrestrial life have a much longer history, in which cosmology (the study of the universe via physics and philosophy) is closely intertwined with faith. One of the stranger consequences of the conflict narrative has been the idea that religious faith and a belief in alien life are mutually exclusive. In 2011, for example, the theologian Ted Peters published a survey of 1300 people, questioned about whether they thought the discovery of extra-terrestrial intelligence would shake their personal belief, the strength of their religion as a whole, or the beliefs of other religions. Amongst respondents who identified as religious, whether Roman Catholics, evangelical Protestants, mainline Protestants, Orthodox Christians, Mormons, Jews, or Buddhists, no threat was seen to their own religious beliefs. However, sixty-nine per cent of those who identified themselves as non-religious believed that the discovery would cause a crisis for world religions.

As this chapter will show, their belief was mistaken; history is filled with pious meditations on the possibility and nature of alien life. The encounter with space is not just a story of technological accomplishment, but also a story of imagination, and as such, it is closely intertwined with theology and the practice of faith. In the following pages, we will explore the ways in which theology and cosmology have considered the existence of other worlds and the possibilities of life beyond the Earth. It will consider the impact that imagined encounters with *intelligent* extra-terrestrial life has had on the development of theology and the role that theology might have in helping humans plan for contact with aliens. It will examine the history of humanity's faith in space, ranging from the relationship between Russian cosmism, space exploration, and the quest for immortality, to the space proselytising of von Braun, who became an evangelical Christian shortly after his arrival in America in 1945. Finally, it will also consider how humans have taken their beliefs with them to space and how they have managed the physical and material practices of faith while in the heavens, so to speak. Rather than looking at an imagined conflict between science and religion, this chapter's focus is on how religious belief and scientific expertise have supported one another as conflict-ridden communities reached for the stars.

A plurality of worlds

Debates about whether or not extra-terrestrial life existed on planets beyond the Earth date back to the sixth and fifth centuries BCE. Anaximander (c.

610–546 BCE) had introduced the concept of the infinite universe, which atomist philosophers such as Democritus and, later, Epicurus developed into what we now know as the theory of the plurality of worlds. In essence, atomist philosophy hypothesised that the world was made up of the infinite combinations of tiny atoms in constant motion, as opposed to being the product of Plato's 'demiurge' in a moment of special creation. If this were true, then it was possible that the Earth was not unique, since there was no reason why the conditions that had led to the Earth's appearance could not be replicated elsewhere. In fact, from this perspective, a limitless number of potential worlds could exist, each holding the possibility of hosting life. Epicurus took Plato's 'principle of plenitude' a step further in relation to this approach, arguing that whatever *could* possibly exist, *would* somewhere, someday, exist.[2] As the first-century BCE philosopher Lucretius noted,

In the Universe, nothing is only one of its kind. In other regions, surely there must be other Earths, other men and other beasts of burden.[3]

Aristotle was opposed to this kind of talk, envisioning a universe that was both unique and finite. Despite the immense influence that Aristotle enjoyed over the rest of natural philosophy, Lucretius' ideas of plenitude and plurality provided fertile grounds for discussions about life beyond Earth for the next thousand years or so. The Aristotelian view was soundly rejected by Islamic theologians such as Fakhr al-Din al-Razi (c.1149–1209), who argued that God could, if God so desired, establish millions of worlds in the void beyond the Earth. Even some of the early Church fathers—Origen, for one—were open to the idea of a plurality of worlds. But, as Chapter 1 demonstrated, it was the enclosed universe of Aristotle, together with its singular Earth, that Thomas Aquinas was able to synthesise with Christian theology in order to create the intellectual context for medieval European scholars.

However, the tension remained: did insisting on the uniqueness of Earth in this way represent an unwarrantable constraint on the omnipotence of God? Again, as Chapter 1 noted, presuming to place limits on divine power was one of the aspects of Aristotelian philosophy that the Bishop of Paris condemned as heresy in 1277. As a result, when the Paris condemnations denied that any such restrictions on God's actions existed, they freed scholastic philosophers like William of Ockham to discuss the possibilities of alternate worlds and to wonder what shape they and their inhabitants might take. What initially looked like religious censorship created territory for freedom of thought. These discussions could take many forms, as the case of Giordano Bruno demonstrates. Bruno was a sixteenth-century

Dominican friar who held a range of beliefs that included the many worlds hypothesis and the existence of extra-terrestrial life. Besides these, he also accepted the Copernican theory of a sun-centred cosmos and the reincarnation of the soul. More dangerously, he denied some core Catholic doctrines including transubstantiation and the divinity of Christ.[4] It is likely that these played a stronger part in his execution by burning than his cosmological beliefs.

By the seventeenth century, speculation about life beyond the Earth was becoming relatively common. In 1608, for example, Johannes Kepler wrote *Somnium*, in which he described a visit to the moon and encounters with gigantic, short-lived reptilian creatures. Wholly unlike anything to be found on Earth, these beings were well-suited to lunar life, which according to Kepler was characterised by profound alterations in temperature and shifting bodies of liquid water.[5] In England and Wales, the clergymen John Wilkins (later Bishop of Chester) and Francis Godwin (Bishop of Llandaff and then of Hereford) both wrote books about visits to the moon. Bernard de Fontenelle used the idea of many worlds to popularise Cartesian and Copernican philosophy in his book, *Conversations on a Plurality of Worlds*. This book, published in 1686, was written in French, not Latin, which meant that it was accessible to a much wider audience. It vividly described the appearance and character of the possible inhabitants of the other planets in the solar system. De Fontenelle moreover argued that each of the stars in the night sky would have its own family of worlds to illuminate—as would the many other stars that were invisible to human sight. There was a theological element to this, as Danish mathematician and philosopher Christiaan Huygens demonstrated: since humans could not see these stars and testify to their creation by God, extra-terrestrial life was necessary to do so. Earthly beings were not the only creatures capable of recognising God's glory.

With the intervention of Thomas Paine, the debate shifted away from the tension between divine omnipotence and human exceptionality, and towards the potential contradiction between God's apparently one-off act of incarnation upon Earth and the possible existence of an infinite number of other worlds. In *The Age of Reason* (1795), Paine used the assumed existence of extra-terrestrial life to pour scorn on traditional Christianity. What was so important about humans, he asked, that God should focus so much time on their problems?

From whence then could arise the solitary and strange conceit that the Almighty, who had millions of worlds equally dependent on his protection, should quit the care of all the rest, and come to die in our world, because, they say, one man and one woman had eaten an apple! . . . Are we to suppose that every world in the boundless creation had an Eve, an apple, a serpent and a redeemer? In this case, the person who is irreverently called the Son of God, and sometimes God himself, would have nothing else to do than to travel from world to world in an endless succession of death, with scarcely a momentary interval of life.[6]

Paine is not arguing here for the non-existence of God. Instead, he is supporting a deist perspective, in which God creates the world, but then refrains from any further interaction or intervention in it. What's intriguing though is that the theologians who objected to his approach did not challenge it by denying the existence of alien life. Instead, people like the Scottish Presbyterian Thomas Chalmers and the Baptist Andrew Fuller argued that humans were indeed the exception because humanity alone had sinned:

Let creation be as extensive as it may, and the number of worlds be multiplied to the utmost boundary to which imagination can reach, there is no proof that any of them, except men and angels, have apostatized from God. If our world be only a small province . . . of God's vast empire, there is reason to hope that it is the only part of it where sin has entered . . . and that the endless myriads of intelligent beings in other worlds are all the hearty friends of virtue, of order, and of God.[7]

Only for humanity did God need to intervene, since only humanity had fallen. In that sense, the Christian account of redemption in combination with the existence of a plurality of worlds beyond the Earth reflected a uniquely human fallibility.

Evolution, immortality, and the noosphere

By the second half of the nineteenth century, influential British figures such as William Whewell and Alfred Russel Wallace were pushing back against the assumption that extra-terrestrial life existed. Their rejection of the idea was based on the accumulating empirical evidence that no other body in the solar system could support life. For Wallace, in particular, who was the co-discoverer with Darwin of evolution by natural selection, the chain of events that had produced intelligent life on Earth was so unlikely that it was implausible it would be reproduced elsewhere.[8] Other nineteenth- and early twentieth-century writers on evolution adopted a different approach. Their perspective was based on the belief that the evolutionary process, while not consciously purposive, clearly seemed directional in its production of increasingly complex organisms on Earth. Assuming that they applied universally, natural laws should encourage the production of intelligent life elsewhere.

One of these scholars was Nikolai Fyodorov, a Russian Orthodox Christian philosopher who believed that the direction of evolution was towards the production of intelligent life. He also argued that the evolutionary trajectory of human intelligence was such as to enable humanity to defeat death. Sometimes cited as one of the predecessors of the transhumanist thinkers discussed in Chapter 7, he believed that increased scientific and technological understanding would soon enable humans to take control of their own evolution. Science would not just allow people to control the natural world and therefore avoid death from natural disasters; it would also soon give humanity the tools to identify and repair bodily decay. These factors would ensure physical immortality for those currently living; eventually, human intelligence would advance science to a point that would enable the material resurrection of those already dead.[9] After the Russian Revolution of 1917, Fyodorov's cosmist ideas were taken up in a more secular form by the Soviet government: the immortality of the soul promised by God was now replaced by the (potential) guarantee by the state of an immortal body.

Space travel was essential to this model of the human future. At the most basic level, if all of humanity's dead ancestors were to be resurrected, then it would not take long for them to consume all of Earth's remaining resources. It was fortunate, therefore, that one of Fyodorov's followers was the Russian space-flight pioneer, Konstantin Tsiolkovsky. Tsiolkovsky is perhaps most famous for his equation which calculates the horizontal speed needed to maintain a minimal Earth orbit. His work also included designs for multi-stage rockets and space stations, as well as addressing the theoretical and practical problems of navigation and heating in space, and designing airlocks that could maintain their seal against a vacuum.[10] Alongside his practical contributions to astronautics, Tsiolkovsky also wrote science fiction, publishing in 1928 a book called *The Will of the Universe*. In it, he outlined his belief in the cosmic power of intelligent human creativity, and his belief that the 'common task of humanity' was to colonise the galaxy. This was an answer to the problem of where to put the resurrected ancestors anticipated by Fyodorov: Tsiolkovsky's rocket research would transport them to other planets. Tsiolkovsky, to whose work Fyodorov's cosmist philosophy was central, was extremely influential within the USSR, and to a limited extent elsewhere. A copy of one of his books was found by Soviet soldiers amongst the possessions left behind by von Braun when he fled to meet the advancing American army.[11]

The belief in an underlying creative pattern for the universe, in which intelligent creatures play a significant part, was further developed by another cosmist, Vladimir Vernadsky. Vernadsky, a geologist and geochemist, was responsible for developing and popularising Eduard Suess' concept of the 'biosphere', which Suess identified as one of three stages in the development of the Earth. The first had been the 'geosphere', or the existence of inanimate matter. The second stage or biosphere began when the geosphere became populated by biological life, which in turn began to reconstitute planetary processes (changing the gaseous balance of the atmosphere, for example). The third stage of planetary development was what Vernadsky identified as the 'noosphere'. For him, this meant the capacity of intelligent life to consciously reshape geological, chemical, and nuclear forces: to reshape the biosphere just as the biosphere had transformed the geosphere. His development of the concept happened in tandem with the work of the controversial palaeontologist and Jesuit priest, Pierre Teilhard de Chardin.

It was Teilhard who had first used the word 'noosphere', in his 1922 book *Cosmogenesis*. While in Paris, Vernadsky attended lectures in which the concept was discussed; Teilhard also attended some of the talks given by Vernadsky. Both were aware of, and drew on, Suess' idea of the biosphere.[12] While Vernadsky's approach was grounded in geology, Teilhard focused on the significance of consciousness and cognition as the crucial components of the third planetary layer of existence. The noosphere was based on the increasingly complex interactions of human minds: as societies came to be more complex, so the noosphere itself became larger and more integrated, eventually culminating in what Teilhard called the 'Omega Point', or the experience of unified planetary consciousness. For Teilhard, this echoed the Christian concept of *Logos*, or as Cardinal Joseph Ratzinger wrote, the spirit of creative reason that produces and permeates the universe.[13] The future Pope Benedict XVI might have approved; the Catholic Church of the 1920s was significantly less keen. Teilhard was banned from teaching and banished to China where the Church hoped he would cause less trouble. His theology, however, continued to resonate with aspects of the work of other Western scientists, from contemporaries such as the ethologist Conway Lloyd Morgan to the twentieth-century polymath James

Lovelock.[14] The Teilhardian view is that life tends to evolve towards complexity, eventually leading to the emergence of intelligent consciousness. If biological laws apply universally, then the same rule will hold true on different planets. These debates—and the work of Teilhard in particular—have proven controversial. Yet, they continue to provide an important resource for cautious consideration of the role of mind in terrestrial evolution as well as for speculation on the nature and likelihood of extra-terrestrial life.

In the early twentieth century, the conceptual work of the Russian cosmists had an immediate practical impact on the development of rocket science within the USSR and, via the influence of von Braun, beyond. Some scholars have argued that the influence of the cosmists meant that the Soviet space programme, unlike that of the US, was a not simply a brutalist demonstration of military power, or the first stages in colonialist expansion into the solar system. Instead, it encapsulated a philosophy that positioned the work of space exploration as an essential part of the evolutionary heritage of humanity. This metaphysical approach contrasted sharply with the 'spiritually impoverished' US space programme.[15] Closer inspection, however, reveals that faith—in God, in technology, and in destiny—was also central to the American efforts to reach the heavens.

Having faith in space

The cultural historian De Witt Douglas Kilgore has produced a fascinating exploration of astrofuturism, which he defines as the Euro-American 'preoccupation with imperial expansion and utopian speculation, [recast] in the elsewhere and *elsewhen* of outer space.'[16] This approach places the inspiration for America's future space programme firmly in the context of the country's history: the new frontiers of space would do for twentieth-century Americans what the Western frontier had done for their nineteenth-century ancestors. Kilgore's analysis focuses on the imaginative context of space exploration: it is not enough simply to show that it is *possible* to go to space, proponents of the mission needed to tell their audience why they should *want* to go there. It is for this reason, he suggests, that people like von Braun and his colleagues Willy Ley (an exiled German) took time to produce popular accounts of the human future in space. They wrote science fiction and worked with Walt Disney to produce the public attraction 'Tomorrowland', creating a widespread cultural expectation that life in space would soon become a reality.

Von Braun's efforts at creating fiction fell flat, but Ley's speculative science, *The Conquest of Space* (1949), illustrated by the painter Chesley Bonestell, became a best seller.[17] It was followed by a highly influential series on the human future in space that appeared in eight instalments in the magazine *Collier's Weekly* from early 1952 to 1954. The overriding question posed by the series was simple: why are we waiting to go to space, when the final frontier is right there, waiting for us? The articles were written by Ley and von Braun, as well as by the respected US astronomer Fred Whipple, the Hungarian-born physicist Joseph Kaplan, the historian Cornelius Ryan, and the law professor Oscar Schachter. The group also included physicist Heinz Haber, who, like von Braun, had been retrieved from military service by the Americans. As with Ley's book, the articles in each issue were lavishly illustrated by the paintings of Chesley Bonestell. Their subjects ranged from human survival in space, how best to select the men [sic] who would go, what it would be like inside the moon ship or the lunar base, and—importantly—what to do when things went wrong. The influence of this series was tremendous.

As Catherine Newell has pointed out, Chesley Bonestell's paintings had a special role in linking the old and the new frontiers. Prior to his involvement with Ley, Bonestell had built a significant reputation through his development of ultra-realistic techniques for presenting imagined astronomic vistas.[18] His pictures enabled the audience to place themselves in in outer space. Long before the Apollo astronaut Bill Anders took the famous 'Earthrise' photo from lunar orbit, Bonestell showed his audiences a similar view through a rocket porthole as they climbed out of Earth's atmosphere. In 1944, *Life* magazine published a series of Bonestell's paintings that hurtled the viewer on a journey towards Saturn. Starting from the most distant Saturnian moon, each successive image represented the sky as seen from the surface of the planet's satellites. Audiences could see Saturn getting larger and larger in each painting until, in the final image, they stood with Bonestell's imaginary artist on the planet's surface, looking up at the planet's rings in the sky above. Two of the paintings also showed tiny space-suited figures exploring the moonscapes. Newell argues that Bonestell's composition and aesthetics reflected an older tradition of American landscape painting, especially the work of artist Thomas Moran.

Moran was the official painter on the Hayden Geological Survey of 1871, sent by the US federal government to explore the lands that eventually became the Yellowstone National Park. Moran's watercolours, together with the photographs taken by his colleague William Henry Jackson, provided the politicians and public back East with compelling visual evidence of the remarkable landscapes and geological marvels that existed in the far West. Moran's work in Yellowstone was so successful that he was invited to join other expeditions organised by the US government to the western territories, including the Rocky Mountains, the Colorado River, and the Grand Canyon, as well as receiving commissions from periodicals and railroads anxious to encourage Easterners to travel West.[19] Some images, such as the cruciform snow fields of *Mountain of the Holy Cross* (1890), seemed to provide visual proof of God's covenant with his newly chosen people. Others, such as the *Grand Canyon of the Yellowstone* (1872), included tiny human figures, overwhelmed by the immensity of the awesome landscape surrounding them. In Moran's paintings, art critics identify resonances with the splendours and terror of Creation, as well as the literal inscription of doctrine on landscape (the tiny snake that can be found in the lower left foreground of the painting, *Chasm of the Colorado*, is another example of this tendency).[20]

Newell argues that Bonestell's paintings need to be seen in this tradition. They are representations of a frontier that is nearly beyond the human imagination yet domesticated through the depiction of human figures within its planet-scapes. Domestication is also achieved through the use of perspectives that ground unsettling visions within mundane human experiences: the unfamiliar lunar landscape framed by the familiar shape of a porthole, for example. In collaboration with von Braun and Ley, Bonestell's paintings provided their lay audiences with a way to vicariously participate in the coming conquest of the final frontier. They represented reasons to have faith in a realisable future in space. Together, these men extracted rockets from the realms of warfare and fantasy and treated them as just another form of travel, albeit a tremendously exciting one. Their promotion of space travel created the groundswell of public support and confident expectation on which the Apollo programme depended.

Above all, Bonestell, Ley, and von Braun connected the prospect of space travel with the historic American belief in manifest destiny. This nineteenth-century creed fuelled the expansion of Euro-American settlers through Yellowstone and towards the Pacific coast. In its space-age articulation, it paralleled the powerful sense of metaphysical purpose of Russian cosmism. It was destiny because it represented great purpose; it was manifest because as the three men had shown, its accomplishment was at hand.

Land at the new frontier was not only a sign of divine blessing to nineteenth-century Americans but also represented a solemn duty incumbent upon them: to spread the blessings of civilisation and good governance across the continent. In the twentieth-century reworking of destiny as duty, the space race was no mere matter of petty rivalry with the Soviet Union. Rather, it was Americans' collective responsibility to take their nation's virtues of democracy, civilisation, and individual freedom before God to the stars. For some, the duty extended beyond nations to humanity or even life at large. Von Braun, who claimed to have experienced meaningful religion for the first time in the US, echoed the cosmists when he remarked:

> If man is Alpha and Omega, then it is profoundly important for religious reasons that he travel to other worlds, other galaxies, for it may be Man's destiny to assure immortality, not only of his race, but even of the life spark.[21]

Science and space colonisation were the path to human salvation, and it was America's privilege to lead the way. The science fiction stories that von Braun and his colleagues created (and consumed) were tales of American enterprise and faith, culminating in the triumph of the true frontiersman. As Gene Rodenberry would put it a decade later when pitching his new series to sceptical TV executives, there would be a 'wagon train to the stars'. *Star Trek* went on to open each episode with its dramatic description of space as 'the final frontier'. The ultimate expression of Euro-American astrofuturism was a Western set amongst the stars.

Gene Rodenberry's vision of a utopian future where people of colour, Russians, and aliens could live in harmony under American leadership tended to gloss over religious belief. However, there are a number of intriguing moments, at least in the original series, in which it is clear that Christianity, or at least some kind of monotheism, is an underlying aspect of life on the *Enterprise*. Encountering the alien god Apollo, Captain Kirk dismisses pantheism with an affirmation of the singular nature of deity: 'Mankind has no need for gods. We find the one quite adequate'.[22] In another episode, Spock wonders why a religion based on sun worship has developed a philosophy of universal brotherhood: Uhura points out that he has misheard what the aliens have been saying. Monitoring their radio broadcasts, she has realised that they are not talking about 'the Sun up in the sky' but discussing instead 'the Son of God'.[23] Viewers are also occasionally allowed to see the *Enterprise*'s chapel which hosts both weddings and funerals. It is not clear to which denomination it belongs, but it does have a stained-glass window.[24] Although the role of religion in fictional space may be debatable, humans going to space in reality have taken their faith with them.

Apollo 8, which launched in December 1968, was the first crewed mission to reach lunar orbit. Expecting considerable public attention, NASA had arranged for the crew to make a live broadcast to the world. NASA had not specified what the astronauts should say; the only guidance to the Misson Commander was that he should 'say something appropriate'. Frank Borman therefore decided to begin 'in the beginning'. So, on Christmas Eve, 1968, while looking out at the view of Earth rising that Bonestell had imagined years earlier, the three crew members took turns to read aloud the first ten verses of the book of Genesis. Thus, they described God's creation of heaven and Earth to the billion or so people—one in four of the planet's population—who were listening. Their broadcast closed with Borman wishing the listeners a Merry Christmas and asking God to bless all the peoples of Earth. This broadcast caused some consternation. The atheist Madalyn Murray O'Hair tried to sue NASA on the grounds that the separation of Church and state meant that astronauts, who were government employees, should not be allowed to pray in public.[25] The Supreme Court, however, rejected the claim on the grounds that it did not have jurisdiction over outer space.

Seven months later, Apollo 11 landed on the lunar surface. During the period of downtime that NASA had scheduled prior to Neil Armstrong's first steps onto the lunar surface, his colleague Buzz Aldrin became the first person to take communion on the Moon. An elder in his Presbyterian Church, Aldrin had received permission to take the sacraments with him into space. Conscious of the controversy caused by the Genesis reading, Aldrin simply requested a moment of silence in which he invited his listeners to pause and give thanks. He then read privately from John's gospel, poured the wine, and ate the bread while Armstrong watched.[26] Aldrin was the first, but not the last, to conduct religious rituals in space. Since then, Catholic astronauts have received Holy Communion on the space shuttle, and the consecrated host has been taken to the International Space Station, which has also seen the celebration of Russian Orthodox Christmas. Jewish and Muslim astronauts have faced particular issues in managing their lived faith, since both require religious observations to be carried out at particular times measured by the position of the sun relative to the horizon. Islam also expects worshippers to adopt particular physical positions (kneeling, facing Mecca) which are hard to manage and maintain in zero gravity. In each case, workarounds have been found, for example, advising that the astronaut should be guided by the appropriate times and dates in their place of origin. Other religious items have also travelled to space. Ed White, one of the astronauts who died in the Apollo 1 launch pad fire, had hoped that he could take a Bible to space with him. In his memory, the Apollo Prayer League commissioned a microfilm of the King James Bible to be taken to—but not left on—the moon. Russian Orthodox cosmonauts have taken icons to space and the Jewish astronaut Jeffrey Hoffman took a dreidel, a small spinning toy associated with the celebration of Hanukkah, on one of his shuttle flights. Seeing him playing with it, Mission Control asked him to explain to his audience what he was doing. After giving an account of the significance of the dreidel and the difficulty of spinning it in zero gravity, Hoffman also showed a miniature menorah he had brought, while reassuring listeners that he had no intention of lighting candles. As well as the Bible and the Torah, the Bhagavad Gita and the Upanishads have been taken to the International Space Station.

Faith has played an important role in supporting those who have gone as far as the Moon and has been considered in relation to conjectural longer journeys. In 2012, the second '100 Year Starship Symposium' was held in Houston by the US Defence Advanced Research Projects to discuss the many challenges that interstellar flight would raise.[27] Star Trek actors Nichelle Nicols (Lt. Uhura) and LeVar Burton (Lt. Geordie LaForge) were in attendance, as were a number of religious leaders: together they debated whether religion should be permitted on the interstellar journey. Some attendees argued that the involvement of the Church would be a key element in making such a mission possible: the resources, organisation, and experience that religious institutions could bring to bear would be invaluable. A significant number of science fiction stories dealing with religion, as it happens, turn on the notion of 'Jesuits in space'.[28] For others, earthly religions had no place in the heavens: fearing sectarian tensions and the growth of fundamentalism, they recommended excluding believers. But as the experience of current astronauts and cosmonauts shows, where people go, they tend to take their faith with them.

Alien theology

The first 'Starship Symposium' (2011) also returned to some well-trodden ground. Harking back to the issue of multi-planetary redemption raised two centuries previously by Thomas Paine, one panel had asked 'Did Jesus die for Klingons, too?' Questions about the need for multiple incarnations and the singularity of the human fall from grace were once again raised. At stake was the issue of whether the encounter with extra-terrestrial intelligence would represent a threat to (Christian) religion, and whether theology could support engagement with aliens.

The first evidence of the existence of a planet outside the solar system, although unrecognised as such at the time, was identified in 1917. The first confirmation of an extrasolar planet came in 1992 and since then more than five thousand such planets have been confirmed.[29] In 2020, a team at the Harvard-Smithsonian centre for Astrophysics identified the first extra-galactic planet in the M51a galaxy.[30] These discoveries certainly pose no threat to theologians who understand God to be the ultimate source of creativity and creation. Nor would the existence of life on these planets be hard to marry with a God that (on Earth) has produced it in such abundant variety. The Christian Bible even includes accounts of non-human intelligence, in the form of angels, and theological debate has never insisted that human anatomy literally mirrors that of God. Why, then, would aliens present a problem? Many modern theologians have engaged extensively and enthusiastically with the question.[31]

Science fiction authors have also explored these themes in some detail. The question of alien incarnation, for example, is examined in Michael Bishop's 'The Gospel According to Gamaliel Crucis' (1983), where human explorers land on a planet whose sentient race is insectoid, gestating thousands of offspring at a time. Multiple messiahs result from God's incarnation on this planet, with one returning with the human explorers to preach the gospel to the inhabitants of planet Earth (naturally including non-human animals in its congregation). Other writers, from C. S. Lewis (*Out of the Silent Planet*, 1938) to Robert Sawyer (*The Neanderthal Parallax*, 2002–2003), have considered the implications of encountering an alien race which acts in perfect morality but has no direct relationship with God. What do these fictional cases suggest about the Christian God's relationship with humans? Is the human species, as Andrew Fuller argued in 1799, the 'one lost sheep' described in the Gospel of St Luke, unique in its need for special help?

Making space together

Despite the stigma and suspicion that developed around 'believing in aliens' in the later twentieth century, the histories of science and theology alike demonstrate that the possibility of life beyond Earth has been constitutive to our understanding of the universe, our planet, and ourselves. Aliens are serious business, and in thinking about them, both science and theology have lessons to learn from each other. The challenging and creative debates currently underway in the developing disciplines of astrobiology and astrotheology both demonstrate this.[32]

Theologians might also play a role in planning for a future human life in space. Some commentators, as discussed earlier, have argued that religion is something that humans should leave behind on Earth but, as the experience of present-day astronauts and cosmonauts indicates, when humans go to space, they take their faith with them. Humans in space are likely to face many of the same challenges and problems that they encounter on Earth: the fair distribution of resources, the management of relationships, and the exercise of compassion.[33] Who will get to go to space? Will it be an escape route, or a prison sentence? Human society has taken different forms in different material and ecological environments; an awareness of a possible world beyond the mundane has played its part in shaping the structures that have made us human. As this chapter has shown, that awareness has also inspired societies to commit scientific, technological, and economic resources to take us to the worlds beyond.

Further reading

Alan P. R. Gregory, *Science Fiction Theology: Beauty and the Transformation of the Sublime* (Waco: Baylor University Press, 2015)

De Witt Douglas Kilgore, *Astrofuturism: Science, Race, and Visions of Utopia in Space* (Philadelphia: University of Pennsylvania Press, 2003)

Lucas Mix, *Life in Space: Astrobiology for Everyone* (Cambridge MA: Harvard University Press, 2009)

Catherine L. Newell, *Destined for the Stars: Faith, the Future, and America's Final Frontier* (Pittsburgh: University of Pittsburgh Press, 2019)

David Wilkinson, *Science, Religion, and the Search for Extra-Terrestrial Intelligence* (Oxford: Oxford University Press, 2013)

Notes

1. 'Man Will Conquer Space Soon!' was the title of the *Collier's Weekly* issue for 22 March 1952, edited by von Braun and others.
2. Arthur Lovejoy, *The Great Chain of Being* (Cambridge, MA: Harvard University Press, 1936).
3. David Wilkinson, *Science, Religion and the Search for Extra-Terrestrial Intelligence* (Oxford: Oxford University Press, 2013), p. 18.

4. Brian Vickers, *Occult and Scientific Mentalities in the Renaissance* (Cambridge: Cambridge University Press, 1984). Bruno was burned at the stake by the Inquisition for heresy in 1600, although it is more likely that he was condemned for his theological doctrine than for his cosmological beliefs.
5. Gale E Christianson, 'Kepler's *Somnium*: Science Fiction and the Renaissance Scientist', *Science Fiction Studies*, 3.1 (1976): 79–90, https://www.jstor.org/stable/4239001.
6. Thomas Paine, *The Age of Reason* (1785), chapter 16 'Application of the Preceding to the System of the Christians', https://standardebooks.org/ebooks/thomas-paine/the-age-of-reason/text/chapter-1-16, accessed 14 February 2025.
7. Andrew Fuller, *The Gospel Its Own Witness*, (New York: Collins, [1799] 1801), p. 189 at https://archive.org/details/gospelitsownwitn00full/page/n13/mode/2up, accessed 14 February 2025; Ralph C. Roper, 'Thomas Paine: Scientist-Religionist', *The Scientific Monthly* 58.2 (1944) 101–11, https://www.jstor.org/stable/18087. C. S. Lewis picked up this idea in his novel *Perelandra* (London: Bodley Head, 1943) which portrays Venus as an inhabited planet in unfallen condition.
8. William Whewell, *Of the Plurality of Worlds* (Cambridge: Cambridge University Press, [1853] 2009); Alfred Russel Wallace, *Man's Place in the Universe* (London: Chapman and Hall, 1903).
9. Boris Gorys, ed., *Russian Cosmism* (Cambridge, MA: MIT Press, 2018).
10. James T. Andrews, *Red Cosmos: K. E. Tsiolkovskii, Grandfather of Soviet Rocketry* (College Station: Texas A&M University Press, 2009).
11. Asif Siddiqi, *Challenge to Apollo: The Soviet Union and the Space Race 1945–1974* (Washington DC: NASA, 2011).
12. Kendall Bailes, *Science and Russian Culture in an Age of Revolutions* (Bloomington: Indiana University Press, 1990).
13. Joseph Cardinal Ratzinger, *Introduction to Christianity* (San Francisco: Ignatius Press, 1968).
14. Conway Lloyd Morgan, *Emergent Evolution* (London: Williams and Norgate, 1923). Although the evolutionary biologist Simon Conway Morris is situated in a more conventional disciplinary trajectory, some aspects of his work bear comparison with Teilhard's big questions; see *Life's Solutions: Inevitable Humans in a Lonely Universe* (Cambridge: Cambridge University Press, 2003).
15. Anton Vidokle and Brian Kuan Wood, 'Foreword' in *Russian Cosmism*, edited by Boris Gorys, (Cambridge, MA: MIT Press, 2018), p. viii.
16. De Witt Douglas Kilgore, *Astrofuturism: Science, Race, and Visions of Utopia in Space* (Philadelphia: University of Pennsylvania Press, 2003), p. 1.
17. Willy Ley and Chesley Bonestell, *The Conquest of Space* (New York: Viking Press, 1949).
18. Realism is an aesthetic convention rather than an exercise of objectivity: no-one can say what a planet 'really' looks like in the absence of a viewer.
19. Catherine L. Newell, *Destined for the Stars: Faith, the Future, and America's Final Frontier* (Pittsburgh: University of Pittsburgh Press, 2019).
20. Joni Louise Kinsey, *Thomas Moran and the Surveying of the American West* (Washington DC: Smithsonian Books, 1992).
21. Catherine L. Newell, *Destined for the Stars: Faith, the Future, and America's Final Frontier* (Pittsburgh: University of Pittsburgh Press, 2019), pp. 199–200.
22. Gene L. Coon and Gilbert Ralston, 'Who Mourns for Adonais?', *Star Trek* Season 2, Episode 2 (1967).
23. Gene Roddenberry and Gene L. Coon, 'Bread and Circuses', *Star Trek* Season 2, Episode 25 (1968).
24. Paul Schneider, 'Balance of Power', *Star Trek* Season 1, Episode 14 (1966); Gene Roddenberry, Judy Burns, and Chet Richards, 'Tholian Web', *Star Trek* Season 3, Episode 9 (1968).
25. Andrew Chaikin, *A Man on the Moon: The Voyages of the Apollo Astronauts* (New York: Viking, 1994).

26. Buzz Aldrin, 'When Buzz Aldrin Took Communion on the Moon', *Guideposts Magazine*, October 1970, https://guideposts.org/positive-living/health-and-wellness/life-advice/finding-life-purpose/guideposts-classics-when-buzz-aldrin-took-communion-on-the-moon/, accessed 14 February 2025

27. Dirk Schulze-Makuch, 'The 100-Year Starship Symposium: A Historic Meeting?' *Astrobiology* 12.1 (2012): 1–2, DOI: 10.1089/ast.2011.0745.

28. Examples include Mary Doria Russell, *The Sparrow* (London: Black Swan, 1996); James Blish, *A Case of Conscience* (New York: Ballantine, 1958); Arthur C. Clarke, 'The Star' (1955); Hilbert van Nydeck Schenck, Jr., 'The Theology of Water' (1982); Damien Broderick, 'The Magi' (1982); and Stephen Baxter, 'Dante Dreams' (1998), in which the Jesuit is a woman.

29. A. Wolszczan and D. A. Frail, 'A Planetary System around the Millisecond Pulsar PSR1257 + 12', *Nature* 355 (1992): 145–7, https://www.nature.com/articles/355145a0.

30. Rosanne Di Stefano, Julia Berndtsson, Ryan Urquhart, Roberto Soria, Vinay L. Kashyap, Theron W. Carmichael, and Nia Imara (2021-10-25). 'A Possible Planet Candidate in an External Galaxy Detected through X-ray Transit', *Nature Astronomy*. 5.12 (2021): 1297–1307, https://www.nature.com/articles/s41550-021-01495-w

31. Ted Peters and Carl S. Helrich, *The Evolution of Terrestrial and Extraterrestrial Life: Where in the World Is God?* (Kitchener, Ont: Pandora Press, 2008); David Wilkinson, *Science, Religion, and the Search for Extraterrestrial Intelligence* (Oxford: Oxford University Press, 2017); Ted Peters, ed., with Martinez Hewlett, Joshua M. Moritz, and Robert John Russell, *Astrotheology* (Eugene: Cascade Books, 2018).

32. Lucas Mix, *Life in Space: Astrobiology for Everyone* (Cambridge MA: Harvard University Press, 2009).

33. Erika Nesfold, *Off-Earth: Ethical Questions and Quandaries for Living in Outer Space* (Cambridge MA: MIT Press, 2023).

5

Genetic modification

Rogue rodents

Cousins of that icon of science, the lab rat, took a turn towards the weird
in the 1990s as experimenters modified the genomes of mice and rabbits.
OncoMouse was an early example, patented in the United States in 1988.
Its tweaked genes made it susceptible to cancer and thus a tool for cancer
research. A fluorescent green rabbit followed in 2000, when artist Eduardo
Kacs claimed to have added DNA from a jellyfish to the genome of the rodent
so that it glowed in a certain light. Although there was debate about whether
Kacs had actually done the science that he claimed, the rabbit's image went
viral, capturing a horrified fascination about the new technology. It complemented photographs that had begun to circulate in 1996, showing a mouse
with an ear growing out of its back. This was not, in fact, the direct result
of genetic modification, but was nevertheless widely taken as emblematic of
the science that had produced OncoMouse, along with a range of genetically
modified (GM) crops that entered the landscape in the 1990s. Control over
the natural world, it seemed, had gone too far.

Concerns mounted about genetic modification as its techniques and technologies moved from experimental to industrial use in the late 1970s and
early 1980s. People who worried about it often spoke about the dangers
of 'playing God'. Others made links to ancient myths of human hubris,
such as Prometheus or the Trojan horse, or to the modern cautionary tales
of *Frankenstein* and *Dr Strangelove*.[1] In the UK, one of the most famous
voices in this chorus was the then heir to the throne, Prince Charles. In
1996, he urged caution in the introduction of genetically modified organisms
(GMOs), stating that 'we have now reached a moral and ethical watershed
beyond which we venture in realms that belong to God and to God alone.'[2]
The invocation of God suggested that the supposed conflict between science
and religion was, once again, the cause of trouble.

In this chapter, we argue that scientists' problematic claim on godlike
qualities happened much further back in the story of science. For many

Science, Religion, and the Human Future. Amanda Rees et al., Oxford University Press. © Amanda Rees,
Franziska E. Kohlt, Tom McLeish, Charlotte Sleigh, David Wilkinson (2025).
DOI: 10.1093/9780191995316.003.0006

people, the success of science stems from the supposed fact that it is an activity uniquely independent of social influence or assumptions. Through the generation of hypotheses and their empirical testing, scientists are supposed to create objective knowledge of the world, untainted by social bias or value-laden judgement.

This chapter will show that controversies relating to GMOs occurred in part because of the problematic assumption that science is independent from society and capable of operating in a value-free and objective fashion. This claim is an outcome of the supposed conflict between science and religion: the notion that science is objective, while religion is subjective. Looked at another way, science's 'view from nowhere', when contrasted with the conflicted views of mere mortals, is remarkably similar to a 'God's eye view'. It follows that scientific research, like God, cannot be held to human constraints—and that its application is the problem of someone other than scientists themselves. Treating science as objective and untainted by social values is unrealistic (more on this in Chapter 11) and is to blame for the often-unfruitful nature of public debate about GMOs. In reality, even 'blue-sky' scientific research encapsulates and celebrates social values, not least the belief that, as the magazine *New Scientist* put it in 2016, 'reason, discovery, and innovation' are essential to social progress.[3] 'Progress', although often not well-defined, is usually linked in some way to a demonstrable increase in the health and wealth of the population (Chapter 3). In this case, agriculture was 'improved', with little evidence, despite the controversies, to suggest that GMOs would cause harm to human or environmental health. However, access to scientific and technological products is not always the result of free choice: sometimes it is imposed and sometimes it can be withheld. Sometimes, often, perhaps, there are unanticipated disadvantages that lie outside immediately testable questions of health. Questions about the ecological and human impacts of the industrialisation of agriculture began at least as early as the mis-named 'green revolution' of the twentieth century.

It is perhaps unsurprising that the hostility of Prince Charles and others was generally treated by proponents of GMO technologies as based in ignorance; mentioning God added fuel to the fire. As we noted in Chapter 3, late twentieth-century debates about public understanding of science, particularly in the United States and United Kingdom, often began from the assumption that if only people knew more about science, then they would be less afraid of it. But public fears in this case related not only to the question of whether a particular technology was judged to be risky: they went to the heart of deeper questions about the kinds of values that were being expressed and mobilised through scientific research. Academic commentators might argue that science and religion occupied non-overlapping magisteria, where expertise would show what *could* be done and ethics would determine if it *should* be so; public debates actively rejected such a tidy division of responsibility.

As this chapter will show, the understanding of scientific responsibility and the effective communication of science in public depend on the kinds of values and narratives that are used to account for the work that scientists are doing. Where different communities perceive these narrative values as isolated from the wider social, ecological, or ethical consequences of that work (or are even accompanied by a derogation of responsibility for them), then public concerns about technoscientific innovation are likely to persist. This is particularly the case where innovations—in this case, GMOs—are linked to wealth generation and where specific individuals or groups are likely to profit disproportionately from the introduction of technology. This chapter will explore the history of responses to the introduction of GMOs. It will show how conflict between narratives stressing the need to manage risk for the benefit of humanity. and narratives emphasising the dangers of interfering with nature, developed within a wider context of assumptions about the relationship between science and morality in society.

Modifying opinions

It was in 1973 that, drawing on Paul Berg's work with recombinant DNA, Herbert Boyer and Stanley Cohen first managed to transfer genes between organisms. The first transgenic animal, a mouse, was created in the following year by Rudolf Jaenisch. The first transgenic tobacco plant was reported almost a decade later. Boyer also created the first genetic engineering company, helping to found Genentech in 1976. By 1980, the US Supreme Court had decided that it was possible to patent GMOs. During this early period, GMOs were largely confined to the laboratory setting: modified mice, for example, were made in order to enable researchers to study cancer, while bacteria that could synthesize human insulin were also in the process of development. By the mid-1980s, the first field trials of GM plants were underway in France and the US. As one might have expected, plants were manipulated to increase their resistance to attacks by disease and insects. However, they were also modified to increase their tolerance to herbicides (weedkiller). These changes facilitated the mechanisation of farming and reduced the need for human labour in the fields. They also created a dependence of GM crops on the same company's brand of herbicide and increased the total amount of herbicide that was sprayed on the crops. Another systemic problem was that the programmed infertility of GM crops meant that farmers could not save seeds for replanting as per traditional practices but were locked into purchasing from the company year-on-year. As a result of these factors, the commercial benefits of GM plants were mostly felt by large-scale farmers, as opposed to smaller, family farms.

Alongside these events came a great deal of speculation about the possibilities of the new technologies. By the early 1980s, for example, many science fiction writers were exploring the potential impact of genetic engineering on society.[4] Most concentrated on the consequences of blurring the boundaries between humans and other species, although some (John Wyndham, for example) had been considering the role that genetically altered plants could play in the human future since the 1950s.[5] By the late twentieth century, the promises made by companies about the future of real GM agriculture began to seem science-fictional. Golden rice would end Vitamin A deficiency. Whiffy Wheat would release an odour to scare off aphids. Crops would decrease phosphate pollution. There would even be perennial crops that would never need replanting. The list of potential benefits that GMOs would bring was extraordinary. Yet, by the early twenty-first century, few of these promises had been realised. Some commentators began to wonder whether 'Big Ag' companies had been engaging in hype.

Back in the early 1990s, there had certainly seemed grounds for optimism. More specific methods for genome editing had been added to earlier techniques, such as the gene 'gun', which enabled more targeted control over genetic manipulation. In 1992, the US Food and Drug Administration had declared that GM foods were 'not inherently dangerous' and—so long as they met the same standards as did foods from other sources—could be sold for human consumption.[6] In 1994, the first GM food, the Flavr Svr tomato, was put on the market in the United States by the company Calgene. The background to this development was the fact that commercially grown tomatoes are usually picked while unripe and complete the ripening process while being transported to market. However, since the 'shelf-life' of the tomato is relatively short, any delays in shipping mean that the crop can rot before it reaches the consumer. Calgene had used a bacterial parasite to infect the Flavr Svr variety with two additional genes intended to slow down the ripening process and delay the onset of rot. While the company hoped that this would also improve the flavour and the consumer experience, the primary focus was on the immediate profitability of industrial food production. In that sense, the GM tomato was a failure: its introduction did not make large-scale harvesting of tomatoes any easier. By 1997, Flavr Svr had been withdrawn from the market, and Calgene had been bought out by the company Monsanto.

Notwithstanding public setbacks such as this, around thirty-five million hectares of GM crops were under commercial cultivation globally by 1998.[7] Crops included soya, maize, tobacco, cotton, and oilseed rape, as well as tomatoes and potatoes. Almost three-quarters of all these crops were planted in the US, with most of the rest in Australia, Canada, or China. In the UK, GM crops were undergoing testing, but none were being sold to the consumer, even though three such items had been approved for the market: a tomato product using something like to the Flavr Svr and herbicide-resistant varieties of soya and maize. A House of Commons report noted that the first introduction of GM materials to the UK diet in 1990—yeast in bread-making and rennet in cheese-making—had not initially caused any public debate or concern.

By the time of the report, however (1998), UK public opinion of GM food was radically shifting. At the beginning of that year, GM tomato paste had been outselling its more expensive traditional competitors by a ratio of two-to-one. By Christmas, food manufacturers and UK supermarkets were falling over each other to declare that their food was genetic modification-free.

What had happened? In the summer of 1998, media attention had focused on the work of Arpad Pusztai, a researcher at the Rowett Research Institute in Aberdeen, Scotland. In a TV interview, Pusztai suggested that feeding GM potatoes to rats caused measurable damage to their health. This generated immediate and immense concern from newspapers and other media. Public anxieties were exacerbated by the fact that there was considerable confusion as to what experiments Pusztai had actually done and the extent to which his claims were based on peer-reviewed research data. But the immediate impact of this—apart from Pusztai's suspension by his employers—was a significant increase in UK public fears about GMOs, or 'Frankenfoods' as some newspapers began to call them.[8] Government officials and scientific institutions tried to draw a clear line between Dr Pusztai's work and that produced by other genetic modification researchers. They condemned the 'science by press release' method of communicating research conclusions, arguing that only results that had been through a recognised process of peer review could or should be trusted. They feared that 'unfounded scare stories' produced by irresponsible media organisations, colliding with the gung-ho pronouncements of publicity-seeking scientists, would combine with public ignorance to deny the UK a role in genetic modification technology development. Members of the public, however, saw another kind of narrative developing from the fact that Pusztai had been sacked. They feared a cover-up.

The media tended focused on the potential health consequences of consuming GM foods. Could modified DNA be bad for human health? It was easy for scientists and multinational PR professionals to rebut these claims, and they did so noisily. Doubters of the technology were explicitly mocked for thinking that ordinary tomatoes did not contain DNA and implicitly so for the ignorant, probably religious perspective that led them to talk about scientists 'playing God'. These staged conflicts took up media oxygen and prevented a focus on the more reasonable ethical arguments, such as the economic consequences for small-scale farmers, the ecological consequences of increased pesticide use, and the possible evolutionary consequences of pesticide- or herbicide-resistance arising amongst natural populations of weed or insect. From the perspective of Big Ag, it was certainly a convenient deflection of debate.

Over the next few years, opposition to GMOs continued to build in Europe, despite official efforts to encourage public acceptance. This hostility was not just verbal: in 1999, the first farm-scale trial of GM crops in Europe was destroyed by protestors. This was the first of many such attacks by people who feared that releasing GMOs into the wider environment could have dire, if unintended, consequences. A relatively clear-cut divide seemed to be appearing, with government, business, and science on one side, and community action groups, NGOs, and private citizens on the other. In an effort to address this cleavage, the UK government instituted one of the biggest efforts ever made to engage the public with scientific innovation. 'GM Nation?', which took place between 2002 and 2003, was made up of a three-pronged approach that saw the science and the economics of GM foods being assessed by experts, while simultaneously creating opportunities for laypeople to contribute to public debates about the safety of the new technology. The UK government committed itself in advance to making sure that laypeople had access to the experts and tools they needed to debate these matters and identifying and addressing their concerns about the commercial use of GM crops. Yet despite spending half a million pounds on hundreds of meetings involving thousands of participants, it wasn't clear what the government wanted. Were they trying to inform the public, to consult them, or to just to make sure that they were participants in the overall process of assessing GM foods? Industry representatives and scientists raised concerns about the methodology of the project, but the results of the exercise were clear. The UK public overwhelmingly rejected the idea of GM crops.[9]

By 2004, other countries including India, Argentina, China, Brazil, and South Africa had small but significant areas of land commercially planted with GM crops. These introductions had similarly been accompanied by public uncertainty and doubt. The 'Monsanto Quit India' campaign launched in the late 1990s alongside crop destruction protests extensive enough that the United States Embassy asked Bengaluru police to step in to protect American commercial interests.[10] The public scholar Vandana Shiva played a key role in focusing the activities of grassroots activist organisations, not just in India, but also in South Africa, where uneasy relationships between government, industry, scientists, and the public were still affected by the legacy of the apartheid era. Protests here were targeted at retail outlets, focusing on the question of adequate labelling and customer choice. Court cases were pursued, to open up the details of genetic modification trial approvals to public scrutiny. The problem of transparency and labelling was a global issue, exacerbated by the fact that the United States did not discriminate between GM and traditionally produced soya and maize, making it impossible to determine what foods were actually derived from GM sources. Court cases also dogged the introduction of genetic modification in Brazil, particularly with respect to disrupting the import of GM foods.[11] A relatively consistent cross-national pattern was emerging. Government, scientists, and industries all insisted that GMOs were not only safe but also necessary to the creation of secure and resilient human food futures. Public groups continued to express disquiet, suspicion, and sometimes outright hostility to their use. Local groups connected with global support, even in the relatively GMO-friendly US, through movements such as the 'March Against Monsanto'.[12] But GM crops were continued to be planted, generating public anger and opposition even in the absence of evidence of long-term harm.[13]

Different regulatory regimes were in play. One key difference between the situation in the United States and Europe was that in the United States, the focus of regulation was on the safety of the product, while in Europe, the concern was with how that product was produced. In the US, where a given *product* could be shown to be equivalent in risk to its non-GM counterpart, there was no reason to require specific labelling. For European agencies, the question of risk assessment focused on showing that the *process* of genetic modification itself was unlikely to cause harm. But the persistence of conflict over decades and continents suggests that this was not just a problem of public communication regarding a novel food technology. To address it, we need to consider the deeper narratives, rhetorical strategies, and metaphors underlying the debate.

One notable technique used by supporters of GMOs was to emphasise that these modifications were just the next step in a much longer history of human engagement with nature. They pointed to the fact that humans had been actively manipulating the breeding of other organisms for around 12,000 years, ever since the shift from hunter-gatherer to agrarian societies that marked the first 'agricultural revolution'. It makes sense, therefore, to look at that history in the context of present-day concerns.

Modification in history, modification of history

In November 2013, the UK's Council for Science and Technology was asked by the Prime Minister to assess the evidence for the risks and benefits of genetic modification technology. It began its written response by referring him to 'the series of continuous experiments that have been running at Rothamsted Research Institute'. These experiments had increased wheat yields eightfold per hectare since their inception in 1843; the expectation was that yields could be increased further.[14] GMOs were thus placed within a genealogy of scientific agricultural improvement. The specific techniques might be new, but all that they were doing was developing work that dated back hundreds, if not thousands, of years. This strategy was consistently used when explaining genetic modification technology to different publics, sometimes with a specific appeal to the figure of Gregor Mendel (the monk who is credited with the discovery of heredity) and sometimes with a more general reference to selective breeding undertaken over the course of human history.

The process of plant and animal domestication probably began around 10,000 to 12,000 years ago with wheat and barley in Mesopotamia, although the ancestors of modern dogs are likely to have lived in close relationships with human groups for even longer.[15] Traditional accounts of domestication were understandably written from a human-centred perspective, focusing on the role of human agency, desires, and needs in this first 'agricultural revolution'. Domestication, on this interpretation, was treated as something that humans did to other species in order to ensure a dependable supply of food, labour power, and other resources. This anthropocentric perspective on prehistory has become considerably more nuanced over recent years. Rather than treating domestication as synonymous with domination, the process is now more often understood as a relationship between species, from which parties can benefit in different ways.[16] Not all species, after all, could be domesticated, despite the best efforts of the people involved—at least, not until the intimate control of genetic modification techniques was developed.

These shifting attitudes towards the relationship between humanity and the natural world have important implications for understanding how the narratives underpinning the GMO debates developed. The 1970s saw researchers start to take this relationship seriously with the emergence of environmental history. Scholars such as William McNeil, Richard Grove, and Roderick Nash began to examine how human attitudes towards nature had changed over time and across cultures.[17] Keith Thomas suggested that during the early modern period, people in England came to value nature precisely because of its uncontrolled and unspoilt state, contrasting the beauty of a forest wilderness with that of a carefully curated garden.[18] Others pointed out that a romanticisation of the landscape occurred only at a point where nature was increasingly considered to be under human control: it was safe to sentimentalise a no-longer-dangerous space. Carolyn Merchant, in her pioneering *Death of Nature* (1980), charted the ways in which metaphors of control, dissection, atomisation, and objectification marked early modern discourse about nature, showing how nature was:

> put in constraint, moulded and made as it were new by art and the hand of man ... Nature takes orders from man and works under his authority.[19]

The European period of the seventeenth to nineteenth centuries, known as the second 'agricultural revolution', was marked by a significant increase in agricultural productivity. A series of changes were implemented in farming, from the adoption of the Chinese plough to the introduction of crop rotation and the organisation of land ownership. A much more systematic and selective approach to plant and animal breeding was also implemented. Desired traits were identified and bred in already-domesticated species. Modern breeds of animals and plants, specialised for their landscape or for the commodities that they produced, were created. This process of artificial selection was deliberately undertaken by individuals seeking to maximise the production value of their crops and flocks: scientific selective breeding was fundamentally intertwined with the capitalist drive to extract profit. Thus, the ideal sheep was defined by the early nineteenth-century agriculturalist Robert Bakewell as the 'best machine for converting herbage into money'.[20]

The intensification of the industrialisation of agriculture that marked the period following the Second World War in Europe accelerated this approach. The experiences of farmers in Europe, North America, and the Global South were all very different—but had in common the drive towards increased mechanisation alongside the use of chemical assistance in the form of herbicides, pesticides, and artificial fertilizers, as well as new breeds of (non-GMO) hybridized seeds. Mexico was a cradle of what became known as the 'green' or 'third agricultural revolution', based on active transfer of technology and expertise with the United States. While many farmers welcomed the bumper crops that the new varieties brought, the key beneficiaries were usually large-scale enterprises. Small farms often could not afford the fertilisers, pesticides, or herbicides required to get the most out of the new seeds. Even if they could, they and others began to feel concerned about the long-term impact of those chemicals on human and environmental health.[21]

In the United Kingdom, the impact of food industrialisation was being felt in ways that would have a direct impact on the genetic modification debate a few years later. By the 1980s, cattle represented the largest sector of UK agriculture, the growth of which depended on the capacity of farmers to supplement grazing with additional feed. Traditionally, agricultural companies had sold farmers products containing bonemeal to feed omnivorous livestock such as pigs and poultry; during the Second World War, this practice extended to herbivorous cows and sheep. Supplementing the feed of calves and young animals in this way meant that they grew and became profitable much more quickly. Understandably, farmers were reluctant to abandon the practice even once the war was over. From their perspective, they were simply adding handfuls of 'protein' to the cows' normal feed. In reality, what this meant was that cows were being fed abattoir waste, including sheep brains and the remains of other cows. The brain disease known as scrapie had been endemic in British sheep since the eighteenth century; in 1986, government scientists diagnosed similar symptoms in cattle. Over the next few years, more and more cattle were found to be suffering from the condition. In the face of increasing public concern, the UK government continued to assure the public that bovine spongiform encephalopathy (BSE), or 'mad cow disease', could not spread to humans who ate beef products. However, in 1994, a new neurological disorder—new-variant Creutzfeldt-Jakob disease—was identified in some patients, and by 1996, the British government accepted that that this condition was linked to eating BSE-infected meat.

Even though few people were eventually diagnosed with the new disease, the affair had a deeply negative impact on public trust in the governance of science in agriculture. Concerns about food quality went back a long way in both North America and Europe; legislation making it an offence to sell food that knowingly endangers health or is adulterated dated from 1860 and 1906 in the UK and the US, respectively. The BSE affair was not the only food crisis to erupt as a result of the industrialisation of the food supply. However, in the ways that government, science, and business were thought to be complicit in ignoring public fears, it encapsulated many of the concerns that would underpin the hostile response to GM foods from 1998 onwards.

To put this history back in the context of the debate surrounding genetic modification, remember that proponents of GM foods had sought to calm critics by asserting that GMOs were part of a long history. They wanted concerned audiences to see genetic modification as just another innovation in the thousand-year process through which humans had remade plant and animal species into more useful and productive forms. The proponents of genetic modification made the point that all they were doing was facilitating natural selection: they were doing what nature did, just in a faster and more targeted way. After all, both processes could produce unforeseen consequences in relation to biodiversity, weed resistance, and crossbreeding. Supporters argued that GMOs, which had been extensively risk-assessed and tested, should be seen within a broader narrative of the imperative to 'feed the world', implicitly placing concerned consumers in the Global North in the position of squeamishly and selfishly denying hungry people access to food security.[22]

When different publics did as they were asked and placed GMOs in the wider history of agricultural science and biotechnology, what they saw instead was a process whereby control and agency had shifted away from individual farmers and consumers and towards corporations and institutions. They saw a clear pattern in which ordinary people had been consistently advised by officials that food was safe and that their concerns were not scientifically justified—only for those concerns to be later shown to be well-founded. In fact, genetic modification could mitigate the dangers of crossbreeding by introducing 'terminator genes' that would prevent the modifications being spread. However, these were unpopular because they would require farmers to buy expensive new seeds every year, to the economic benefit of manufacturers. Rightly or wrongly, critics perceived a situation in which worries about the broader consequences of particular innovations, ecological or socio-economic, were rejected in favour of a short-term (and potentially short-sighted) focus on immediate risk assessment. Such concerns were not only expressed by consumers in the Global North. Publics in Mexico, Brazil, and India were also deeply wary of the impact and introduction of GM foods. Despite the very different cultural and economic experiences of these communities, the themes on which they drew to account for their concerns resonated strongly with those identified in the European and North American context.

From outcome to governance

In response to a funding call from the John Templeton Foundation, an interdisciplinary group of academics met in Durham University (UK) in 2011 to discuss the global response to GM crops. They did not focus on why people ought to accept GM crops; instead, they began from the context of local concerns and cultures, asking why people *had not* found these crops acceptable.[23] The *GM Futuros* project focused on three case studies based in Mexico, Brazil, and India and was undertaken in collaboration with local partners. These partners conducted in-country ethnographic field research with farmers, researchers, consumers, and those involved in food preparation and sale in selected areas. This combination of participant observation and interviews meant that researchers could examine how the prospect and experience of GM crops was embedded in specific contexts and practices, both in private (for example, in laboratories) and in public (for example, at food festivals). Each case study focused on responses to a particular crop: maize in Mexico, soya in Brazil, and cotton in India. The ethnographic research was further contextualised by systematic work with key stakeholders from government, NGOs, consumer groups, manufacturer representatives, and groups representing the interests of particular communities within the area. In-country empirical work was followed by workshops at Durham University, as well as a day's discussion at the Royal Society in London, to review and compare the analyses of the wide range of agro-socio-ecosystems that the teams had studied.

Unsurprisingly, the project found that global responses to GMOs were notably different on key points. For example, in Mexico, the tremendous cultural significance of maize meant that any modification to that crop was particularly problematic, while Brazilian communities treated GM soya as a symbol of wider inequalities relating to landownership and US hegemony. In India, GM cotton was especially challenging because of the historical link between cotton and Gandhi's call for Indian self-sufficiency and the agency of the poor. But project researchers also found that the plurality of voices to which they listened emphasised some common themes. The most important of these was that lay communities framed the debate about GMOs in much wider terms than did government officials, scientists, or corporations. This was not a question of risk assessment; it was instead a debate about the socio-economic and ethical responsibilities of science and innovative technologies.

The questions of trust and transparency were critical. The extent to which people felt that the introduction of GM crops was done at the behest of powerful actors was particularly significant. These actors included national governments, corporations, businesses, and international agencies that could not be held accountable to local populations in any satisfactory way. Where these national and transnational actors stressed the beneficial nature of their engagement (lower costs, higher profits for farmers), their motives (feed the world's hungry), and their reliability (based on sound science), these claims were not accepted as reliable. In particular, 'science' was seen as under the control of business and government, rather than an independent actor in itself.

This resonated strongly with studies of technology governance in the Global North, perhaps most emphatically with the experiences of the Cumbrian sheep farmers in the UK studied by Brian Wynne in the 1990s. In the aftermath of the Chernobyl nuclear disaster of 1986, severe restrictions had been placed on the movement of radiation-contaminated sheep from the hill farms of north-western England. These restrictions, introduced in the immediate aftermath of the catastrophe in Ukraine, were only expected to last for a few months. In practice, they persisted for many years; the last restrictions on animal movements on the farms were not removed until 2012. The farms most affected were close to one of Britain's own nuclear power plants at Sellafield in Cumbria. Sellafield, then called Windscale, had been the site of the world's worst nuclear disaster prior to Chernobyl, when one of the plant's nuclear piles caught fire and burned for days in 1957. The relevance of this fact did not escape the notice of local farmers, most of whom had lived in the area for generations. Scientists and government denied any connection between the earlier accident and the continuing radioactivity of sheep after Chernobyl, but as a local farmer put it:

> They talk about these things coming from Russia, but it's surely no coincidence that it's gathered around Sellafield. They must think that everyone is completely stupid.[24]

Based on their own past and present experience of scientific advice and government guidance, British hill farmers rejected what they saw as a collective official effort to evade responsibility for environmental and economic harm in the aftermath of nuclear disaster. The communities studied by the *GM Futuros* project showed similar suspicions when the objective voice of 'science' was heard to offer advice that seemed to coincide with the perceived interests of corporations and governments.

This might be seen as further proof that the authority and success of science depends on its separation from society. But communities were clear that their concerns about the introduction of innovative technology were not just related to discomfort with technocratic governance, but to their vivid awareness that the *consequences* of technoscientific innovation were experienced socially. The reason why controversy continued in this context was that participants framed the debate according to different parameters. Scientists and governments were focused on risk-based assessments, asking questions about the safety of GM foods for human and environmental health, and evaluating the reliability of the answers they were getting. While the wider community certainly had concerns about risk, their questions also concerned disparity and fairness, focusing on the limits of scientific knowledge and the likelihood of unintended consequences. The community feared that the costs and benefits of innovation were not going to be borne equally, either within human societies or across the more-than-human world.

Narratives that the European DEEPEN project identified while researching public concerns with the ethics of emergent nanotechnologies reflected similar concerns.[25] The philosopher Jean-Pierre Dupuy suggested that some of these framing narratives were very old, relying on the mobilisation of ancient myths. These included warnings about being careful what one wishes for, the dangers of opening Pandora's box, or the consequences of interfering with nature. Others were more explicitly modern: only the rich will benefit, and unwelcome information will be covered up.[26]

Scientists, holding to the belief that science itself is unaffected by social values, are typically prepared only to discuss the possible impact of their discoveries. But as these studies show, ordinary people often want to discuss how politics, money, and other values affect the actual research. There is little public space for the conversations that they want and need to have.

Communities can support scientific research and technological innovation because of the benefits it can bring; increasingly, they also ask that these activities be undertaken with an eye to their wider consequences. As researchers such as Jack Stilgoe and Philip Macnaghten have argued, this involves moving away from an assessment of risk (outcomes) and towards the governance of innovation (development).[27] Dialogue between different groups and stakeholders provides essential, but not sufficient, conditions for improvement. As Macnaghten and colleagues have suggested, dialogue needs to be accompanied by willingness on the part of scientists to accept a moral responsibility regarding the way that their knowledge is utilised, anticipating the kinds of futures and consequences that innovation might bring. Research that can change the world cannot be treated as just another day at the office.

The need to anticipate consequences as part of research and development requires an adeptness with science-fictional thinking. As we noted earlier, many of the promises of genetic modification technology have always seemed to exist more in the realm of science fiction than fact, a characteristic shared with the debates about the development of autonomous vehicles, nanotechnology, RNA medicine, and many other innovations. As we'll see in Chapter 7, fiction has had a powerful, and not always helpful, role to play in shaping responses to the prospect of artificial intelligence. When it comes to imagining socio-technical futures, science fiction is consumed by many more people than read the reports of the Royal Society. It enables its readers to empathise with individuals and communities in ways that go beyond their own lived experience. Many of the fears for GMOs were summed up in the notion of 'Frankenfoods'—monstrous creations that would escape human control and turn on their makers. The real tragedy of Mary Shelley's story, however, was not that Dr Frankenstein tried to play God by creating life. Rather, it lay in his failure to be a good father to the being he had created. Frankenstein failed to nurture and love his creation. As the philosopher Bruno Latour has pointed out, we need to love our technology; to imagine its future with, or without, a proper parent, rather than adopting an instrumental approach towards it. We need to imagine with care and compassion what it might be or become when it grows up and leaves its laboratory home.[28]

Further reading

R. Douglas Hurt, *The Green Revolution in the Global South: Science, Politics and Unintended Consequences* (Tuscaloosa: University of Alabama Press, 2020)
Phil Macnaghten and Susana Carro-Ripalda, *Governing Agricultural Sustainability: Global Lessons from GM Crops* (London: Routledge, 2015)
Gemma Milne, *Smoke and Mirrors: How Hype Obscures the Future and How to See Past It* (London: Robinson, 2020)

Notes

1. Philip Ball, *Modern Myths: Adventures in the Machinery of the Popular Imagination* (Chicago: University of Chicago Press, 2021).
2. HRH The Prince of Wales, 'Speech on the 50th Anniversary of the Soil Association: The 1996 Lady Eve Balfour Memorial Lecture', 19 September 1996, https://www.royal.uk/clarencehouse/speech/speech-hrh-prince-wales-50th-anniversary-soil-association-1996-lady-eve-balfour-memorial, accessed 14 February 2025.
3. New Scientist, 'The World Needs Scientific Values More Than Ever', *New Scientist*, 16 November 2016, https://www.newscientist.com/article/mg23231004-400-the-world-needs-scientific-values-more-than-ever/, accessed 14 February 2025.
4. Helen Parker, *Biological Themes in Modern Science Fiction* (Ann Arbor: UMI Research Press, 1984).
5. Adam Stock, 'The Blind Logic of Plants: Enlightenment and Evolution in John Wyndham's *The Day of the Triffids*', *Science Fiction Studies*, 42.3 (2015): 433–57, https://doi.org/10.5621/sciefictstud.42.3.0433.
6. Food and Drug Administration, 'Science and History of GMOs and Other Food Modification Processes', 2023, https://www.fda.gov/food/agricultural-biotechnology/science-and-history-gmos-and-other-food-modification-processes, accessed 10 February 2024.
7. Select Committee on Science and Technology [UK Parliament], 'First Report', 12 May 1999, https://publications.parliament.uk/pa/cm199899/cmselect/cmsctech/286/28602.htm, accessed 14 February 2025.
8. James Randerson, 'Arpad Pusztai: Biological Divide', *The Guardian*, 15 January 2008, https://www.theguardian.com/education/2008/jan/15/academicexperts.highered.ucationprofile accessed 14 February 2025.
9. J. Giles, 'UK Public Opposes Government on Transgenic Crops', *Nature* 424 (2003), 33 (2003), https://www.nature.com/articles/425331b.
10. Ronald J. Herring, 'Why did "Operation Cremate Monsanto" Fail? Science and Class in India's Great Terminator-Technology Hoax', *Critical Asian Studies* 38.4 (2006): 467–496, https://doi.org/10.1080/14672710601073010.
11. Ian Scoones, 'Contentious Politics, Contentious Knowledges: Mobilising against GM Crops in India, South Africa and Brazil', *IDS [Institute of Development Studies] working paper* 256 (November 2005), https://assets.publishing.service.gov.uk/media/57a08c6be5274a27b20011bf/wp256.pdf, accessed 15 February 2025.

12. John S. Dryzek, Robert E. Goodin, Aviezer Tucker, and Bernard Reber, 'Promethean Elites Encounter Precautionary Publics: The Case of GM Foods', *Science, Technology and Human Values* 34.3 (2008): 263–288, https://doi.org/10.1177/0162243907310297.

13. Matt Reynolds, 'As Kenya's Crops Fail, a Fight over GMOs Rages', *Wired Magazine*, 3 March 2023, https://www.wired.co.uk/article/kenya-gmo-approval, accessed 15 February 2025.

14. Mark Walport and Nancy Rothwell [Council for Science and Technology], Letter to David Cameron, UK Council for Science and Technology, 21 November 2013, https://assets.publishing.service.gov.uk/media/5a7b7fc2e5274a7202e17921/cst-14-634-gm-technologies.pdf, accessed 15 February 2025.

15. Pat Shipman, *Our Oldest Companions: The Story of the First Dogs* (Cambridge MA: Harvard University Press, 2021).

16. Greger Larsen et al., 'Current Perspectives and the Future of Domestication Studies', *PNAS* 111.17 (2014): 6139–46, https://doi.org/10.1073/pnas.1323964111.

17. See, for example, Roderick Nash, *Wilderness and the American Mind* (New Haven: Yale University Press, 1967); Richard Grove, *Green Imperialism* (Cambridge: Cambridge University Press, 1994); William McNeil, *The Human Condition* (Princeton: Princeton University Press, 1980).

18. Keith Thomas, *Man and the Natural World* (London: Allen Lane, 1983).

19. Carolyn Merchant, *The Death of Nature* (New York: Harper and Row, 1980), p. 170.

20. Jason Hribal, 'Animals Are Part of the Working Class: A Challenge to Labour History', *Labor History* 44.4 (2003): 435–453, p. 437, https://doi.org/10.1080/0023656032000170069.

21. R. Douglas Hurt, *The Green Revolution in the Global South: Science, Politics and Unintended Consequences* (Tuscaloosa: University of Alabama Press, 2020).

22. Royal Society *Reaping the Benefits: Science and the Sustainable Intensification of Global Agriculture*, 2009, https://royalsociety.org/news-resources/publications/2009/reaping-benefits, accessed 15 February 2025.

23. Phil Macnaghten and Susana Carro-Ripalda, *Governing Agricultural Sustainability: Global Lessons from GM Crops* (London: Routledge, 2015).

24. Brian Wynne, 'Misunderstood Misunderstandings: Social Identities and The Public Uptake of Science', *Public Understanding of Science* 1 (1992): 281–304, p. 288, https://doi.org/10.1088/0963-6625/1/3/004.

25. S. Davies, P. Macnaghten, and M. Kearnes, eds., *Reconfiguring Responsibility: Deepening Debate on Nanotechnology* (Durham: Durham University, 2009), https://durham-repository.worktribe.com/output/1609054, accessed 15 February 2025.

26. Jean-Pierre Dupuy, 'The Narratology of Lay Ethics', *NanoEthics* 4.2 (2010): 153–170, https://doi.org/10.1007/s11569-010-0097-4.

27. Jack Stilgoe, Richard Owen, and Phil Macnaghten, 'Developing a Framework for Responsible Innovation', in *The Ethics of Nanotechnology, Geoengineering, and Clean Energy*, edited by Andrew Maynard and Jack Stilgoe (London: Routledge, 2020).

28. Bruno Latour, *Aramis, or The Love of Technology* (Cambridge MA: Harvard University Press, 1996).

6

Climate change

A matter of belief

Watching Al Gore's film *An Inconvenient Truth* (2006) was, for many people, the moment that crystallised the reality of the Earth's changing climate. What had previously existed as a mixture of concerns about pollution and species loss now came to focus on a specific and over-arching threat: the heating planet.[1] Within a decade of its release, *An Inconvenient Truth* was reckoned as the eleventh-highest grossing documentary of all time in the US.[2] This figure underplayed the film's true audience reach; the DVD was also screened in many churches, universities, and other civic venues to non-paying audiences.

Following the film's success, the fossil fuel industry quickly redoubled its efforts to cast doubt upon the evidence and the arguments that Gore had deployed within it.[3] Climate change, they suggested, was a false belief. There was no need to believe it; there was no reason to believe it; there were many reasons *not* to believe it. On the other side of the fight, environmentalists redoubled their assertions that climate change was indeed a matter of fact and a rational belief.

How should this difference in belief be handled in the public square? The BBC, Britain's national broadcaster, held for some time to a notorious policy that mandated the inclusion of so-called climate sceptics as a counterbalance to those who believed in anthropogenic (human-made) warming, even though the scientists claimed that the evidence was settled in favour of the latter position. On 13 February 2014, with severe flooding affecting much of the southwest of England, BBC Radio 4's morning news programme asked whether climate change was a factor in such extreme weather events.[4] The comments of Sir Brian Hoskins, Professor of Meteorology at the University of Reading and a leading climate scientist, were 'balanced' by remarks from Nigel Lawson. Lord Lawson, a politician with no scientific credentials, was sceptical about climate change and had links with organisations that lobbied

Science, Religion, and the Human Future. Amanda Rees et al., Oxford University Press. © Amanda Rees, Franziska E. Kohlt, Tom McLeish, Charlotte Sleigh, David Wilkinson (2025).
DOI: 10.1093/9780191995316.003.0007

against policies to limit greenhouse gases. After this debacle, a year-long inquiry by Parliament's Science and Technology Select Committee criticised BBC News for continuing to 'make mistakes in their coverage of climate science by giving opinions and scientific fact the same weight'.[5]

It was the time of the 'Web 2.0', with the rapid growth of blogs, social networking, and video-sharing sites (notably YouTube). These media facilitated the rapid dissemination of information and misinformation intended to convert people to one tribe or the other—the believers or the sceptics. People began to experience difficulty in personal and professional relationships when confronted with people who did, or did not, believe—whichever view it was that conflicted with their own.

Social scientists—sociologists, social psychologists, anthropologists—launched a host of surveys to discover who believed in climate change and who did not, and what factors predicted the camp into which people would fall.[6] Women were found to have greater knowledge of the matter than men, and more concern about it.[7] Localised perceptions and expectations about the weather caused the climate to be perceived as either more or less stable than the real rate of change.[8] Researchers also covered questions of how to provoke a change in belief; confronting people with fear-inducing facts about the effects of heating did not, they found, enhance belief.[9] And so on. Before the turn of the millennium, believers and non-believers were not greatly divided by politics. A Gallup Poll of 1998 found that, in the US, Democrats and Republicans differed by only one percentage point on the question of whether they believed that anthropogenic climate change was real and underway.

One survey specifically questioned the effect of *An Inconvenient Truth* upon belief, finding that, in its wake, the percentage of Americans accepting the effects of human beings upon the climate rose from forty-one to fifty per cent.[10] The apparently positive effect that Gore's film had on belief was countered three years later by the 'Climategate' campaign (2009). Hackers stole the private communications of climate scientists based at the University of East Anglia (UK) and republished them in selective and misrepresented form so as to smear their credibility. Right-wing commentators and media seized upon the evidence of a climate-change 'hoax' with glee. The Climategate campaign likely exacerbated what was already a trend towards scepticism: the Pew Research Center reported a drop in belief in the evidence for global warming, from seventy-one per cent in 2008 to fifty-seven per cent in 2009. By 2011, believers and non-believers had clustered around political poles, with Democrats 'twice as likely as Republicans to be concerned about climate change and to think it is already impacting the planet.[11]

Throughout this time, and indeed beforehand, fossil fuel companies employed lobbyists to cast active doubt upon the believability of climate change. The strategy began as far back as the 1970s and continues to the present day; Naomi Oreskes and Erik Conway have investigated its shocking history in their book *Merchants of Doubt*. The strategy was successful, in part, because of an unrealistic model of science that once again has roots in the conflict thesis. Karl Popper's account of scientific method, popular amongst scientists, states that science can never prove anything but only ever attempts to disprove things. This model of science was congenial to anti-religious scientists, who were able to caricature faith as rooted in unfalsifiable belief. When extended to anthropogenic climate change, the model made final proof an unachievable goal and encouraged a constant unpicking of individual data which could potentially falsify it.[12]

In short, the public discourse of the 2000s and the 2010s centred upon a question of belief. Do you believe in climate change, people asked one another? The answer to that question put people in one tribe or the other: the believers or the non-believers. People were assigned into in-groups and out-groups according to their belief. Believers might shake their heads as they commented of a colleague or a figure in the news: *did you know he doesn't believe in climate change?*

The situation has striking parallels with the way that people, in the Western world, have tended to talk about religion for the past few decades: *Are you a believer?* It sounds like the kind of question that might be posed at an evangelist's rally. 'To a person who's a true believer', said the disgraced televangelist Jim Bakker, 'if you die, you know you're going to heaven'.[13] Belief was core to his type of Christian evangelism and to the subject matter of many other preachers. Prominent atheists (see Chapter 10) enjoyed contrasting the apparently baseless grounds for being a religious believer against the supposedly rational grounds for believing in science. 'You can't convince a believer of anything', remarked Carl Sagan, 'for their belief is not based on evidence, it's based on a deep-seated need to believe'. This, then, was the immediate cultural backdrop for the 'belief' debate in climate science. It did not help matters that evangelical Christians in the US were quickly co-opted into climate scepticism, apparently confirming to atheist and agnostic scientists that these people were prepared to ignore scientific evidence in order to maintain their belief. (The prominent US climate scientist Katharine Hayhoe, who is an evangelical Christian, has worked hard and in the face of considerable abuse to argue that the science is not antithetical to their shared theology.)[14]

Meanwhile, the unexamined assumption for the climate believers was that once everyone had become a believer like them, then the problem would quickly be resolved. Many sociological and socio-psychological studies blurred the distinction between belief and individual action, judging the former on the basis of the latter. Within many of these studies, a person was implicitly assessed to have a real belief in climate change if they were taking steps to reduce their carbon footprint.[15] On a social or systemic level, the assumption was rarely made explicit, though it seemed like common sense. For how could a sane world proceed otherwise? It seemed obvious: once the belief had been accepted, then it was a small step of rationality to develop policies and behaviours that would save everyone from misery and death. This, as we are realising in the 2020s, is not the case.

The belief/non-belief debate that raged for the first twenty years of the millennium was a false start for climate change, an unfortunate import from the conflict-thesis approach to science. Talking about belief was no more useful for climate change than it ever was for religion. Belief did not, as it transpired, lead to action. What was needed was faith. Faith is not the same thing as belief, nor is it equivalent to knowledge. Belief is a thin thing, an individual mental condition that is hard either to create or to change deliberately. Faith, on the other hand, is a much more complex orientation towards a kind of imagined future reality. While faith can be underpinned by knowledge, knowing the facts about a situation is insufficient to produce change within it. Action only happens when individuals and communities have faith that their actions will create that imagined future. This framework of belief, knowledge, faith and imagination explains a great deal about the difficult history of climate change science and politics, as we shall see in the following section. The development of knowledge and belief was dependent upon the facilitation of imagination; the connection between belief and action even more so. As with our other scientific examples (space, food, AI, disease), faith turns out to be an ineradicable component of how we imagine and construct human futures.

A history of climate science belief

The first person to notice the heating effects of carbon dioxide was Eunice Newton Foote.[16] In a series of experiments published in 1856, Foote put jars containing different gases on a windowsill. She discovered that the jars containing carbon dioxide heated up more than those containing normal air when they were exposed to the sun's rays. (The physicist John Tyndall, whom we last met in Chapter 3 giving an address to the BAAS in Belfast, did very similar experiments some three years later; historians debate whether this was a coincidence or whether he knew about, or should have known about, her precedent-setting work.) Despite the fact that Foote (and Tyndall) knew that the factories of Victorian Britain were pumping out carbon dioxide, they had not 'discovered' the greenhouse effect. They simply did not make the connection. A laboratory demonstration just did not feel like the real world. Perhaps they thought about it as a theoretical possibility, but nothing they said or wrote indicated their grasp of the possibility that human activities were already at a scale to cause these effects on a global scale. They had no imaginative sense of the Earth as a changing system.

A next step in the retrospective history of climate change occurred at the turn of the twentieth century. In Sweden, Svante Arrhenius used physical chemistry to estimate what degree of warming would actually occur on Earth. He looked back in time to the ice ages and found data to suggest that temperature changes were actually happening. Arrhenius, then, had a deep historical sensibility that Foote lacked, and had the imagination to make the leap to possible change in the present. A newspaper article of 1902, modestly entitled 'Hint to coal consumers', presented Arrhenius's findings to its readers. It warned that in a few tens of thousands of years, the Earth would be 'baked in a temperature close to the boiling point'. Yet even this did not constitute a 'discovery' of climate change. The laconic title of the newspaper article indicates something of the ambivalence with which this discovery was received. Yes, it was true, but at the same time, it was a possibility too fantastical in which to have faith. Tens of thousands of years seemed like a long time (even though one would presumably reach 'uncomfortably hot' long before the final boiling point). It was true, presumably, but also meaningless.

During the twentieth century, scientific discoveries about climate came thick and fast. The United States and other cold war powers began to gather copious amounts of data in connection with military interests, such as submarine voyaging. Anomalous weather data indicated warming trends. Yet the imagination of the Global North—where this information was principally recorded and disseminated—was focused on its own concerns, such as the ability to make rain. As a rule, people in the Global North suffered more from the cold than from the heat. A little bit of warming seemed like a good thing. Once again, the particularities of imagination framed the factual belief in a way that prevented a grasp of the global issue.[17]

In 1958, the superstar director Frank Capra brought a film about the warming world to American primetime TV. (A similar film, fronted by Prince Philip, was screened in the United Kingdom.) Entitled *Unchained Goddess*, Capra's film spoke of the 'extremely dangerous' possibility that 'man may be unwittingly changing the world's climate through the waste products of civilisation'. It identified automobiles and factories as the culprits, explained the greenhouse effect, and projected sea-level rise due to ice-melt. But here, the film took a turn into whimsical fantasy, much like the article of 1902, with a cartoon animation showing tourists enjoying an aquatic tour of underwater Miami. The science was in place, but the imaginative grasp was, once again, lacking. It just seemed too distant from everyday experience. The title, perhaps, said it all: could the Earth really be a 'goddess' who would rise up and take revenge? The truth was far-fetched in the extreme.

The US president commissioned a report to get the facts in 1965; by the 1970s, they were indisputable. A CIA (Central Intelligence Agency) report of 1974 affirmed the geopolitical risks predicted by the data, against a backdrop of famines around the world caused by changing climate patterns. The projections delivered to the US president in 1979 were basically the same as we have today. And yet—still yet—there was no faith in the matter. For one thing, the problem did not sit in the chronological frame for thinking about the near future within which politicians and business leaders operate. Scientists reported that '[w]hen you go to Washington and tell them that the CO_2 will double in 50 years and will have major impacts on the planet, [they say] come back in 49 years'.[18] Moreover, it was not clear what kind of a problem this was: was it technical, economic, or political? Was it a problem about sea levels or about famine? About geopolitical instability? Over-population? Factories?

The fossil fuel lobbyists further impoverished the imaginative framework for thinking about and responding to climate change. A report responding to scientific findings around Climate Change in the 1970s, led by economist Thomas Schelling, set out the thoughts 'that climate sceptics would echo for the next three decades'.[19] For one thing, future prediction was fraught with uncertainty and was no basis for present action. It was constantly liable to challenge via the falsification of some particular detail. Besides this, said the sceptics, climate change was not intrinsically bad. Humans would simply need to 'adapt' and migrate (as they always had done) in response to shifting climate zones. From the 2010s onwards, sceptics began to modify their arguments, moving away from outright denial and into more a subtle limitation of imagination. Climate scientist Katharine Hayhoe observes that old-fashioned denial has, as of 2024, evolved and diversified into five distinct responses: (1) 'not real', (2) 'not us', (3) 'not bad', (4) 'too hard/costly to fix', or (5) 'too late'.[20]

By the 1970s, the consensus belief in climate change rested on a number of factors. These included: a scientific model (that is, the work done by Foote); confidence to apply the model to human activity (the work of Arrhenius); data of different kinds (Cold War science); and knowing that getting hotter is bad (getting out of the Global North perspective). And yet this belief still, right now in the present, fails to provoke action. It is not a belief that goes very deep.

The preconditions that would enable deeper belief, or action, have proved harder to establish. One of them is the imaginative framework to contain the diverse data that we are seeing: famine and climate change, in the 1970s, were two different ways of looking at unfolding events. We are still stuck with the same question that stymied the US presidents of the 1970s: is this a problem primarily for economists, businesses, technologists, scientists, politicians, or someone else?

Constraints of imagination and emotion work against faith-enabling frameworks. Capra's sci-fi tourists, floating above a subaqueous city, are a good example. The viewer of the 1958 film is invited to identify with the future *viewer* of the drowned city, not its drowned or exiled inhabitants. The viewer is safely insulated from reality both in the present (it hasn't happened yet) and future (it won't happen to you; in fact, you will be titillated by looking at it). There is no imagined continuity between the 'us' of the present day and the future 'us'—or rather 'them'—who will suffer. Rather, the imagined continuity is between 'us' in the present and a future 'us' that is safe, consuming with curiosity the remnants of the past. The film creates, in other words, an imaginative 'us and them' to act as a lens for thinking about climate breakdown. 'We' are the people of observation, of so-called scientific objectivity, and not the people who experience, or are subject to, climate change.

It is easier for the world's wealthy to identify with people like themselves in the future than with the present-day world's poor. Current discussions about climate change still often focus upon the future, disregarding the suffering that is already happening. A report calculated that 676,000 people were killed by climate change between 1991 and 2022, with developing countries suffering seventy-nine per cent of deaths and ninety-seven per cent of the people affected by the impacts of weather extremes.[21] It requires imagination for a British person, part of the global minority, to look out of the window on an ordinary, grey day, neither warm nor cold, and to contemplate the reality of climate extremes happening elsewhere. In much the same way, a smoker believes that their habit causes lung cancer and yet they do not have faith that it is worth resisting *this* cigarette, in this moment.

Another imaginative gap concerns the rate of change. Apocalyptic film and fiction present global change as instantaneous and overwhelming. *The Day After Tomorrow* (2004) is a climate-related example, featuring tsunamis that course between New York's skyscrapers. The reality is less cinematic, with flood zones flowing more and more often, and a little more each time. Each time the crops, the homes, and the infrastructure are damaged a little more severely and take longer to repair. This is the part of the story that is missing from Capra's depiction of Miami, which has no narrative of change, only a before and after. New Orleans has already witnessed the in-between moment in Hurricane Katrina (2005). The city's poorest residents, mostly African Americans, bore the brunt of overtopped levees and lives lost. Their destroyed properties were often uninsured and their livelihoods were knocked yet a little further back, awaiting the next flood.

In a similar vein, discussion about climate change at the beginning of the twenty-first century suggested that it was a go/no-go event, something that could be stopped if everyone acted correctly. Scientists now take pains to emphasise that this is not the case. Global heating is locked in by emissions to date; the question is how well it can be curbed. Every fraction of a degree matters. Malevolent agents take advantage of this on/off imaginative framework to argue that it is 'too late' to mitigate global heating, with some success.

So, then, to believe in climate change, in addition to all the scientific factors listed above, much more is required—factors that are arguably still not in place. These factors are well-summarised by the word 'imagination'; a power to take what is known rationally and to embed it in our perception of what lies beyond our doorsteps, reaching out across the globe in the present and into the world's future.[22]

Faith, imagination, and wicked problems

The early twentieth-century philosopher Charles Saunders Peirce provides a description of faith that works well for those who are struggling to extricate themselves from the mutually exclusive belief systems that are the legacy of the conflict thesis. (By contrast, many Indigenous belief systems do not separate the 'material' and the 'spiritual'; for these people, thinking about climate change may in some ways be easier.[23]) Peirce is interested not in abstract belief (*does it exist or not?*), but in 'a living, practical belief' (*is it worth my assent?*). He argues that if a person muses on the idea of God's possible existence, they will come to an idea of God that is compelling—so compelling that, if real, it would demand assent. God is defined as the thing which, if real, would justify one's faith and is therefore logically worthy of commitment.[24] Peirce uses imagination as the pivot for faith, just as one might do for a better climate future.[25]

Once we, like Peirce, take imaginative factors into account as necessary to the condition of belief, the definition or nature of belief is transformed. It would be more appropriate to talk about 'faith' than 'belief' in these circumstances. Faith is the ability to see a different future, to orientate oneself and one's community towards it. Maria Stuttaford has commented that 'if Christians have one transferable skill to put on their CVs, it is the imagination to see a better world'.[26] And as for Christians and their 'Kingdom of God', so it is for many other religions. Belief is merely a matter of rearranging one's mental furniture, the content of one's brain. Faith is more like belief that is framed as active assent: the construction of an imagined future (whether tragic or ameliorated) that demands engagement in the present. We saw something similar at work in Chapter 4, in the imagination of space travel and colonisation.

Imagination is the key into transforming climate change from belief to real, experiential fact, and a prompt for change. By imagining a better climate future, the responsibilities and actions that would produce this future come into focus. This is more productive than starting with an undefined 'problem'. The undefined-problem approach to climate change is an example of what is known as a wicked problem: one that has too many dimensions to consider all at once.[27] Address one aspect, and the 'but what abouts?' that one has temporarily put to one side for the sake of focus multiply to the point of inundation. Interestingly, though, a wicked problem is also sometimes defined as a problem which can only be understood *after* one has solved it. This might sound ridiculous, but that is exactly what the imaginative approach can achieve.

The philosopher Bruno Latour has helped to clear away some of the conflict-thesis legacy that has turned climate change into a wicked problem. Latour, who died in 2022, saw very clearly how social justice and ecological crisis are always interwoven.[28] One of his great insights was that the division of the world into 'natural' and 'human' things is both false and harmful.

Latour realised that many things which we treat as natural are only made real for us by a complex network of equipment, people, and conceptual frameworks. Thus, for example, we can't touch carbon dioxide. It became real for us through developments in isolating the gas, stabilising the means of testing for it—and so on through a long history, partially described above, of measuring it in the atmosphere. The sciences that made the technologies that produced the carbon dioxide are the same sciences that made the technologies to measure it and, eventually, to stabilise it as 'the thing that causes global heating'.[29] Carbon dioxide is, to use Latour's term, a 'hybrid'. It is a combination of a molecular form and of the technologies that enable us to 'see' it, as well as of the political discourse that names it as a global enemy.

Latour argues that we forget or erase the networks that enabled us to produce hybrids such as carbon dioxide and then start acting as though they were natural entities, like a stone we have stumbled across. The problem is that we then start asking how we can grapple with these entities, forgetting that they were constructed by humans in the first place. The relevance of Latour's argument to climate change is that it was not carbon dioxide that caused the problem, but rather a host of behaviours, structures, and institutions. It's not carbon dioxide that is to blame but *certain people making carbon dioxide in certain ways*. It is much more effective, therefore, to begin the mitigation with these people and the processes than by painstakingly purifying the molecule from the networks that produced it and then trying to figure out what networks can be mobilised to un-produce it. Trying to un-produce carbon dioxide via geo-engineering or carbon capture is actively harmful because it colludes with other forms of social injustice. Rather than diminishing the unsustainable exploitation of Earth's resources, it just pushes the impact elsewhere.

Imagination for positive climate futures will most likely resist the distinction between human and nature. This imagination requires a form of active empathy that goes against the traditional Western, scientific mindset of pure (or 'godlike') objectivity. We are not, like Capra's viewers, observers of climate change, but participants in it, both as causal agents and as those who will suffer in it. It is a form of imagination that resists the divisive two-sidedness that goes along with conflict thinking. It is also, for Christians, a better theology for science than 'godlike' objectivity, since the Christian God is a God who suffers in compassionate communion with creation.

As many Indigenous people knew long, long before Latour began to write, the necessary imagination does not merely consider human agents on their own but as kin with other species and with distant and future generations. In a deep sense, we *are* those other species, those future generations; they are our kin. This kind of interconnected and intergenerational thinking is better practised by many Indigenous religious systems.[30] The book *What Kind of Ancestor Do You Want to Be?*, which is rooted in this wisdom, starts with the premise—startling for most Western people—that all people are progenitors of the future world.[31] Even those who do not have biological children of their own are creating a world for future generations. When humans think about their identity in this radically expanded timeframe, they might just imagine success and validity in different terms than immediate economic success. Scholars such as John Kapya Kaoma refer to the ethics that emerge from this form of imagination as eco-social justice or biotarianism.[32]

Such an expansive and interconnected climate imagination requires emotional or spiritual resilience. It requires the ability to face up to the human and multispecies death that is already locked in by historic and recent emissions. This, in turn, requires the imagining person to allow the death of others to confirm their own vulnerability, their own mortality.[33] It requires the ability to face up, also, to the death of the enlightenment fantasy that nature could be controlled by rational means. It is, for some, a hard blow to human hubris: to the egotistical myth of individual autonomy.

Imagination is also required to see that the future could be different. This is the form of imagination that can envision what a better world would look like, one that has tackled inequalities and has enshrined nature as something that is shared and sacred. It is the means to cut the Gordian knot of climate change as wicked problem. Only once this imagination is exercised can strategies for mitigation and adaptation become thinkable, let alone achievable. This imagination, by its nature, answers the question with which 1970s US presidents struggled: namely, what manner of problem is this and what is needed to solve it? An imagined new transport system, for example, will depend upon the successful enrolment of politicians, economics, technologies, behaviours, and ecological sensitivities.

The sociologist Stephen Hughes has conducted research exploring the imaginative components of climate science, focusing on community resistance to proposed fracking in the Republic of Ireland which resulted in legislation (2017) outlawing the practice. Hughes concludes that the protesters' achievement represented a 'startling success' of imagination.[34] Drawing on the sociological concept of socio-technical imaginaries (something very close to what is defined in this chapter as 'faith'), Hughes explains how a diverse collective of 'farmers, fishermen, artists, professionals, parents' mobilised its resistance. He notes that their ability to engage their imagination was what enabled them to transcend their political and cultural differences. Together, they imagined what the future that was being imposed upon them might look like, and how it could be different. They became imaginatively focused on what would constitute a 'good life'. People suddenly felt the fragility of the land that they farmed or walked. Knowing that a massive gas installation might be pitched there very soon, they became acutely aware of its value and how much they wanted it to remain. What they had previously taken for granted was now crystallised in gratitude and protectiveness. The emotional quality of the response was captured in the name of the key activist collective: 'Love Leitrim'. In turn, the invocation of love created a space where participants were moved to connect their resistance to fracking with other stories. Pop-up monuments celebrating the landscape were adopted to stand for personal stories of bereavement and loss; people reflected on their shared history of the Troubles and its resonances with the contemporary crisis.

One of Hughes' most interesting findings echoes the points made above about kinship. He argues that a crucial piece of the imaginative work that resulted in the defeat of the frackers was a change in the conceptualisation of self. The activists, he argues, developed a power to imagine the self as part of a collective that would live better in the future. This is a powerful shift. From an imaginative reconceptualisation of the self as living in interdependence, the ability to act as such in the present is released. It is an embodiment of Peircian faith.

There are many other instances of people mobilising imagination to address climate change through the collective creation of better worlds, ranging from the ultra-local (community allotments, repair cafes) to the national or international (Sunrise Movement). One relatively well-known example is the international transition town movement, which aims to help people plan on a local, grass-roots basis to increase their self-sufficiency in relation to energy use and food production. One interesting thing about the movement is that it was originally developed to combat the fear that oil would run out, rather than the damaging consequences of burning fossil fuels. It turned out that a project that was imaginatively focused on social justice had the necessary flexibility to accommodate a shift in scientific emphasis. By contrast, starting by imagining technological solutions tends to produce a less adaptable response if the science evolves.

It's easy to attack these kinds of schemes by saying 'it won't work everywhere' or 'it may solve *this*, but it doesn't solve *that*'. None of these solutions is complete or global in its reach and in a sense that is the whole point of imaginative climate faith. Climate faith is incompatible with the enlightenment fantasy of 'rational', totalising solutions. Like all forms of faith, it commits itself to the hope that seeds of hope will lead to a harvest of bigger future change, through unknown and perhaps unexpected means.

One strong argument for imagination-fuelled climate faith comes of pointing out that *all* climate science involves a faith of one kind or another—much of it unsatisfactory. There is a particularly common but naive form of climate imagination which is known as techno-solutionism. This can involve quite wild schemes of geo-engineering, such as scattering tiny mirrors in space. Carbon capture is spoken about more often: filters that are supposed to scrub carbon dioxide out of the air, so that the unwanted gas can be pumped underground. As of 2024, carbon capture accounts for about 0.1 per cent of emissions, mostly—and invidiously—associated with fossil fuel extraction. No one has yet come up with an economically scalable version of the technology, still less one that can remove the more diffuse emissions that occur away from extraction sites and power stations. Yet despite the distant prospect of carbon capture ever becoming a viable technology at scale, it is built into the net-zero plans of governments around the world, an entirely fictional component of a climate faith that insists on preserving the present systems of consumption. Techno-solutionism pivots upon faith in a future where nothing needs to change for the world's wealthy communities.

Fiction writers, as one might expect, are well able to develop and articulate climate imagination. Kim Stanley Robinson is one example: his novel *Ministry for the Future* (2020) thinks its way into the best possible future available from the starting conditions we have in the present. Although many aspects of his solutions are technical in nature (including economics and social sciences), they are rooted in a concept that respects the principle of interdependence: the Ministry is specifically charged with advocacy on behalf of future generations. It answers the question 'what kind of ancestor do you want to be?' via concrete, governmental means. These modes of imagination are also coming to fruition in real life. Around the world, young people are taking governments to court, arguing that they are failing to protect their rights to live in the future, and demanding tougher climate action in the present. Many are succeeding. Researcher Joana Setzer comments on the power of such cases to further drive the power of collectivised imagination: 'It's not just about the case itself but about creating momentum and bringing people together'.[35]

Some of the most profound thinking about climate faith appears in the writings of science fiction author Octavia Butler. *Parable of the Sower* (1993) and *Parable of the Talents* (1998) explore a future (beginning in 2024) in which climate change and far-right politics have devastated life in the US. Lauren Olamina, central character of the first novel, leaves California and gathers up a band of fellow refugees as she journeys north. Eventually, she sets up a self-sufficient, eco-theological community based on her improvised religion known as Earthseed. The principle of interconnection is at the heart of this faith: 'All that you touch, you Change. All that you Change, Changes you'. Over the past few years, activist communities have gathered around these texts to inspire and inform their connected work in racial and ecological justice.[36]

Beyond belief

In late 2016, American scientists were alarmed by the election of a president who announced his disbelief in climate change, declaring it to be a hoax. By the spring of 2017, their movement of resistance, 'march[es] for science', had gone global. The marches hinted at something that was perhaps more like faith rather than belief in scientific work—an imagination that incorporated at least human interconnection. Banners proclaimed 'science: serving the common good' and 'science: speaking truth to power'. These were new ways of speaking about science, a social and even prophetic vision that was notably missing from the modest falsification goals of the mid twentieth century. It remains to be seen how boldly and effectively scientists will grapple with the economic, social, and political hybrids that constitute climate change—whether they can step into the culturally edgy notion that science is a matter not just of belief, but of faith.

Turning this around, we could also think of faith as a kind of science in its orientation towards practice. As we saw in Chapter 5 (and will see again in Chapter 11), scientific research cannot be cleanly separated from its real-world applications. In fact, it is often inspired by implicit or explicit ideas about its eventual impact, whether immediate or speculative. Like science, then, faith creates a preferred vision of the future ('the kingdom of God') which its practitioners attempt to bring about through their liturgical, moral, and spiritual efforts. The history of utopian writing and practice illustrates the frequently hazy boundary between science and faith in this regard.[37] Monastic traditions, in particular, are undergoing something of a renaissance as climate-positive forms of life (not necessarily co-housed), developing practices and technologies of living that will yield an imagined future of compassion and sustainability.

Reflecting on the connection between Latour's Catholic faith and his account of science, scholar Donna Haraway observed that '[b]elief is toxic to both science and religion'.[38] Her comment underlines the theme of this chapter: that focusing on belief offers a poor account of how people relate to climate change science. It is a product of conflict-based thinking that, by creating a binary choice between 'belief' in science or religion, ends up mis-describing both. 'Belief' yields a poor account of the nearly two centuries of interconnected tests and technologies that have cemented anthropogenic climate change as fact. It is too weak, too subjective, as though belief were just a personal psychological orientation. Those facts about the warming planet were established in the world, not in somebody's state of mind. 'Belief' is also nowhere near enough to describe the kind of science that will draw participants into creating a flourishing and just future. Science is always a kind of faith; the challenge and the invitation is to make sure that it is fuelled by a life-giving imagination.

Further reading

Alice Bell, *Our Greatest Experiment: A History of the Climate Crisis* (London: Bloomsbury, 2021)
John Hausdoerffer et al., eds., *What Kind of Ancestor Do You Want to Be?* (Chicago: University of Chicago Press, 2021)

Notes

1. Although this chapter focuses on climate change, there are reckoned to be nine intersecting 'planetary boundaries', seven of which are beyond critical points. Climate change is the best-known but by no means the only ecological crisis in progress. Katherine Richardson et al., 'Earth beyond Six of Nine Planetary Boundaries', *Science Advances* 9.37 (2023), unpaginated, DOI: 10.1126/sciadv.adh2458.
2. John Cook, 'Ten Years On: How Al Gore's *An Inconvenient Truth* Made Its Mark', *The Conversation*, 30 May 2016, https://theconversation.com/ten-years-on-how-al-gores-an-inconvenient-truth-made-its-mark-59387, accessed 1 September 2024.
3. This history is described in Naomi Oreskes and Erik Conway, *Merchants of Doubt: How a Handful of Scientists Obscured the Truth on Issues from Tobacco Smoke to Global Warming* (London: Bloomsbury, 2010), pp. 169–215. The DeSmog Blog tracks individuals and organisations casting doubt on climate science: https://www.desmog.com/, accessed 1 September 2024.

4. https://www.theguardian.com/environment/blog/2014/mar/26/bbc-failing-robust-debate-climate-change, accessed 18 February 2025.

5. https://publications.parliament.uk/pa/cm201314/cmselect/cmsctech/254/254.pdf, accessed 18 February 2025. The media practice of 'bothsidesism' has been argued to damage the public's belief in scientific consensus. In one study, researchers presented people with two positions about climate change as equally valid perspectives, even though one side was based on scientific agreement and the other was not. They found that in such a scenario, people tended to have less confidence in science and less concern about climate change. M. N. Imundo and D. N. Rapp, 'When Fairness is Flawed: Effects of False Balance Reporting and Weight-of-Evidence Statements on Beliefs and Perceptions of Climate Change', *Journal of Applied Research in Memory and Cognition*, 11.2 (2022): 258–271, https://doi.org/10.1016/j.jarmac.2021.10.002.

6. Matthew J. Hornsey, Emily A. Harris, Paul G. Bain, and Kelly S. Fielding, 'Meta-Analyses of the Determinants and Outcomes of Belief in Climate Change', *Nature Climate Change* 6 (2016), 622–626 (2016), https://doi.org/10.1038/nclimate2943; Kate Sambrook, Emmanouil Konstantinidis, Sally Russell, and Yasmina Okan, 'The Role of Personal Experience and Prior Beliefs in Shaping Climate Change Perceptions: A Narrative Review', *Frontiers in Psychology* 12 (2021), unpaginated, https://doi.org/10.3389/fpsyg.2021.669911.

7. Aaron M. McCright, 'The Effects of Gender on Climate Change Knowledge and Concern in the American Public', *Population and Environment* 32 (2010): 66–87, https://doi.org/10.1007/s11111-010-0113-1.

8. Janet Swim et al., *Psychology and Global Climate Change: Addressing a Multi-Faceted Phenomenon and Set of Challenges. A Report by the American Psychological Association's Task Force on the Interface between Psychology and Global Climate Change*, (Washington DC: American Psychological Association, 2009), pp; 33–34, https://www.apa.org/science/about/publications/climate-change, accessed 16 February 2025.

9. Saffron O'Neill and Sophie Nicholson-Cole, '"Fear Won't Do It": Promoting Positive Engagement with Climate Change through Visual and Iconic Representations', *Science Communication* 30.3 (2009): 355–379, https://doi.org/10.1177/1075547008329201.

10. John Cook, 'Ten Years On: How Al Gore's *An Inconvenient Truth* Made Its Mark', *The Conversation*, 30 May 2016, https://theconversation.com/ten-years-on-how-al-gores-an-inconvenient-truth-made-its-mark-59387, accessed 1 September 2024.

11. Gallup, 'Environment', https://news.gallup.com/poll/1615/environment.aspx, accessed 16 February 2025.

12. Charlotte Sleigh, 'The Abuses of Popper', *Aeon*, 16 February 2021, https://aeon.co/essays/how-popperian-falsification-enabled-the-rise-of-neoliberalism, accessed 1 September 2024.

13. The Best of Interviews with Jim Bakker and Tammy Faye Messner, 'CNN Larry King Weekend', 24 June 2001, www.cnn.com, accessed 1 September 2024.

14. Global Weirding with Katharine Hayhoe, 'The Bible Doesn't Talk about Climate Change, Right?', https://youtu.be/SpjL_otLq6Y?si=uXMnDOE6BArnn9Kf, accessed 24 February 2025.

15. See, for example, Elke U. Weber, 'What Shapes Perceptions of Climate Change?' *Wiley Interdisciplinary Reviews: Climate Change* 1.3 (2010): 332–342, https://doi.org/10.1002/wcc.41.

16. This section is greatly indebted to Alice Bell's brilliant book *Our Greatest Experiment: A History of the Climate Crisis* (London: Bloomsbury, 2021). See also Sarah Dry, *Waters of the World: The Story of the Scientists Who Unraveled the Mysteries of Our Oceans, Atmosphere, and Ice Sheets and Made the Planet Whole* (Chicago: University of Chicago Press, 2021).

17. Charlotte Sleigh, 'The Big Freeze', *Wellcome Stories*, 2022, https://wellcomecollection.org/articles/YhOFNRMAADOEby1l, accessed 16 February 2025.

18. Naomi Oreskes and Erik Conway, *Merchants of Doubt: How a Handful of Scientists Obscured the Truth on Issues from Tobacco Smoke to Global Warming* (London: Bloomsbury, 2010), p. 173.

19. Naomi Oreskes and Erik Conway, *Merchants of Doubt: How a Handful of Scientists Obscured the Truth on Issues from Tobacco Smoke to Global Warming* (London: Bloomsbury, 2010), pp. 174-175.

20. Katharine Hayhoe, 12 April 2024, X, https://x.com/KHayhoe/status/1778783856 499233194, accessed 27 June 2024.

21. Oxfam, '189 Million People Per Year Affected by Extreme Weather in Developing Countries as Rich Countries Stall on Paying Climate Impact Costs', 24 October 2022, https://www.oxfam.org.uk/media/press-releases/189-million-people-per-year-affected-by-extreme-weather-in-developing-countries-as-rich-countries-stall-on-paying-climate-impact-costs/, accessed 1 September 2024.

22. The sociological concept of 'the imaginary' links imagination to social groups and their shared traditions and practices. Within Science and Technology Studies, 'sociotechnical imaginaries' are often discussed, state-level frameworks for conceiving of and governing technologies. In this chapter, we use imagination in a slightly looser sense that reflects faith's dual nature as personal practice and a practice framed by tradition and church.

23. Charlotte Sleigh, 'Liberation Science-ology: Indigenous and Postcolonial Frameworks for Science and Religion', in *Science and Religion: Approaches from Science and Technology Studies*, edited by Zara Thokozani Kamwendo (London: Palgrave Macmillan, 2024), https://doi.org/10.1007/978-3-031-66387-1_11.

24. Charlotte Sleigh, 'In the Beginning Was the Wort: A New Natural Theology of Meaning for Ecological Catastrophe', *Anglican Theological Review* 105.4 (2023): 390–408, https://doi.org/10.1177/00033286231202208.

25. Peirce calls this method of imagining the endpoint, then reasoning backward, 'abduction'. The most famous account is given in Umberto Eco, *The Sign of Three: Dupin, Holmes, Peirce* (Bloomington: Indiana University Press, 1983).

26. Private communication.

27. For a historical and critical review of climate change and wicked problems, see Ernst M. Conradie, 'Why, Exactly, is Climate Change a Wicked Problem?' *Philosophia Reformata* 85.2 (2020): 226–242, https://www.jstor.org/stable/27073910.

28. For an approachable overview of Latour's life and work, see his obituary in the *Guardian* newspaper, 10 October 2022, https://www.theguardian.com/world/2022/oct/10/bruno-latour-obituary, accessed 16 February 2025.

29. There are, of course, other gases such as methane that cause global heating, but CO_2 has been made into the focus of the phenomenon.

30. Patricia Dudu Ngwena, 'African eco-theology: land, ecology, and indigenous wisdom in the works of Samson Gitau, Kapya Kaoma and Jesse Mugambi', Master of Theology, University of South Africa (2020), DOI:10.13140/RG.2.2.14495.18086.

31. John Hausdoerffer et al., eds., *What Kind of Ancestor Do You Want to Be?* (Chicago: University of Chicago Press, 2021).

32. Kapya J. Kaoma, *God's Family, God's Earth: Christian Ecological Ethics of Ubuntu* (Zomba, Malawi: Kachere, 2014); Kapya J. Kaoma, 'Killing Our Children's Children: From Humanitarianism to Biotarianism in Earth-Theology', *Journal of Theology for Southern Africa* 156 (2016): 71–89, https://www.academia.edu/30578096/Killing_Our_Childrens_Children_From_Humanitarianism_to_Biotarianism_in_Earth_Theology, accessed 16 February 2025.

33. Robert Bringhurst and Jan Zwicky, *Learning to Die: Wisdom in the Age of Climate Crisis* (Regina: University of Regina Press, 2018); Roy Scranton, *Learning to Die in the Anthropocene: Reflections on the End of a Civilization* (San Francisco: City Lights Publishers, 2015).

34. Stephen Hughes, 'Hearts and Minds: The Technopolitical Role of Affect in Sociotechnical Imaginaries', *Social Studies of Science* 54.6 (2024): 907–930, https://doi.org/10.1177/03063127241257489. Hughes uses the language of emotion to explore how Irish citizens engaged their imaginations, but as we implicitly argue here, the language of faith might work equally well.
35. Charlotte Kellaway with Joana Setzer, 'Why are Climate Action Cases Rising?', 19 March 2024, https://www.lse.ac.uk/research/research-for-the-world/sustainability/climate-legislation-and-litigation, accessed 1 September 2024.
36. *Octavia's Parables*, https://www.readingoctavia.com/about, accessed 1 September 2024; Adrienne M. Brown and Walidah Imarisha, eds., *Octavia's Brood: Science Fiction Stories from Social Justice Movements* (Chico, California: AK Press, 2015).
37. Gregory Claeys, *Utopianism for a Dying Planet: Life after Consumerism* (Princeton: Princeton University Press, 2022).
38. Donna Haraway, 'Present to Bruno, from Donna', *Social Studies of Science* 53.2 (2023): 165–168, p. 165, doi/10.1177/03063127231157395165.

7

Artificial intelligence

The Turing test

In 1950, Alan Turing published an influential paper on 'Computing Machinery and Intelligence', which began with the provocative question: 'Can machines think?'[1] Having posed the query, Turing then spent some time setting out the terms of his debate, giving careful consideration to the question of how to define 'machine' so as to exclude 'men born in the usual manner' from the competition. In his paper, Turing swiftly rejects the premise of his initial question, deeming it ultimately unanswerable. Instead, he proposes an experiment, based on what he called the 'imitation game'. This game is played by three people: a man (A), a woman (B), and an interrogator (C) who may be male or female and who cannot see either of the other two players. The aim of the game for C is to determine which of A or B is female by asking them questions, to which they give written responses. A's job is to confuse the interrogator by pretending to be female: B's is to help the interrogator as far as she can. Turing then asked: if a machine were to take the part of A, would this make it easier or harder for C to decide which of the other two players was female?

The Turing test, in the form that we now know it, is presented in a very different way. In its most common incarnation, the test simply asks whether C can tell which of A or B is human and which machine. Issues of gender, or of deliberate efforts to mislead, have been pared away from the experiment. Even more significantly, perhaps, the version of the test which has passed into public culture frames it as an ordinary conversation between participants. Turing, however, positioned the trial as a series of questions or challenges that the machine might be asked to negotiate. A, he suggested, might be asked to write a sonnet, to do arithmetic, to solve a chess problem—even to describe the length of their hair.

Despite these shifts in emphasis, Turing's paper sketches out a number of issues that are still at stake in discussions of artificial intelligence (AI). These questions also resonate with the wider question of the relationship between

Science, Religion, and the Human Future. Amanda Rees et al., Oxford University Press. © Amanda Rees, Franziska E. Kohlt, Tom McLeish, Charlotte Sleigh, David Wilkinson (2025).
DOI: 10.1093/9780191995316.003.0008

science and religion. The most obvious point relates to definition: the questions of what makes a human *human* is one that belongs, at least in part, to theology. And what is thinking? What do we mean by 'intelligence' and how does that concept relate to ideas about instinct, learning, or cognition? What does mechanical intelligence mean? Turing's paper is also framed by a series of other conventions that continue to structure debates about AI in the West, most significantly, the tendency to think in terms of dualisms. Key contrasts here include machine/human, artificial/natural, specific/general, man/woman, hope/fear—even, drawing on Turing's example of hair length, mind/body. One could also add servant/master or saved/damned to the list of AI-related binaries, reflecting some of the fears about the consequences of artificial minds that have developed in fact and fiction since Turing's day.[2] Last of all, we again confront the supposed binary of science and religion. The structure of thinking about AI draws on a dualism that is deeply rooted in Western culture.

Turing included an explicit consideration of spiritual objections to thinking machines in his original essay. For him, these objections centred around the function of a soul: thinking, he contended, is what the soul does; and since only humans have souls, only humans can think. His counter to this possible objection rested on the fact that it represented an unwarrantable restriction on divine power. An omnipotent God, he argued, could surely confer a soul on any being, whether machine or elephant. This power would presumably be exercised 'in conjunction with a mutation which provided the elephant with an appropriately improved brain to minister to the needs of that soul'.[3] For Turing, the issue of the soul was a cover for the real reason for theological unwillingness to accept the possibility of machine thinking. For Turing, this more fundamental objection lay in the commitment to the idea that humanity was superior to the rest of creation, precisely because only humanity can think. In this way, intelligence is both the cause and the empirical proof of human superiority over creation: it is intelligence that enables humanity to manipulate the natural world. Thinking machines are dangerous because they would threaten this commanding position.

Turing's paper contributed to the debates about the capacity of science and technology to transform humanity's understanding of itself. This chapter will explore the key themes he sketched out, particularly with respect to the idea of what is meant by 'intelligence' and how that inflects the understanding of who, or what, is 'human'. Drawing on fact and fiction, it will consider where the concept of 'intelligence' came from and how it has been understood in relation to human communities and to the kinds of intelligence that can be artificially created. It will consider some of the global narratives and consequences that surround AI innovations, placing them in their historical and economic context, before turning to look at the role of religion in mediating debates around the role of AI in human futures. The chapter will close by considering the ways in which AI debates encourage us to reconfigure our understanding of the relationship between faith and science. Theological or faith-based metaphors are shown to play a big part in the debate on the future of intelligence. Rather than science and religion being at war, we find that their frames of meaning are conflated; AI researchers draw on theological frameworks whether they know it or not. The supposed existence of a gap between science and religion means that this conflation is rarely noticed, with the result that valuable insights are being lost. To critique AI effectively, we must critique its implicit theology.

What is intelligence?

Part of the problem when it comes to discussing AI is the way in which the idea of 'intelligence' itself has been developed and used over the past century and a half. In public discourse, intelligence is often treated as an objective, identifiable, and measurable characteristic that can be used to distinguish between different categories of being. Aristotle's concept of reason was used for a long time to distinguish between humans and other animals. In the nineteenth and early twentieth centuries, intelligence was operationalised to create divisions between humans that placed white European males at the apex of nature.

The concept of intelligence was developed in its modern form by the nineteenth-century biologist George Romanes. Expanding on the work of Charles Darwin, Romanes took a comparative approach to analysing the cognitive and psychological capacities of different kinds of animals. He did so by establishing a hierarchical ranking from least to most advanced. From his work, the modern idea of 'intelligence' arose, soon coming to be broadly to be understood as the capacity for pattern-recognition, problem-solving, and reasoning. It could be used as a yardstick to classify and order populations.[4]

Thanks to the work of the eugenicist Francis Galton, among others, intelligence came to be regarded as an innate capacity that was differentially distributed among human individuals and populations.[5] Some people, it seemed, were just smarter than others and, as a result, were rewarded with more access to economic, social, and political resources. This was a conclusion which was to have particularly repellent consequences in the late nineteenth and early twentieth century period of intensive European imperial exploitation. Today, the practice of measuring and ordering through formal examination performance persists in many countries. This is supposed to produce a system of meritocracy, though results show that these assessments are rarely linked with any substantial change to social mobility.[6] In the face of all this, it should be plain that the idea of 'intelligence' is not necessarily an objective, external index of performance. Instead, it needs to be understood as a concept that arose in specific historical circumstances and as one that has often been used to naturalise extant social hierarchies based on class, race, or gender.[7]

Despite these problems, the assumption that 'intelligence' could be operationalised through the assessment of reasoning, numerical, and linguistic problem-solving skills has been, and remains, a very significant resource for conceptualising machine intelligence. In 1960, a decade after Turing wondered how we might demonstrate machine thinking, Marvin Minsky published a review of then-current scientific efforts to create AI. As Turing had done, Minsky caveated his review by noting that no generally agreed definition of intelligence existed: for his purposes, he chose to treat it as the capacity to solve problems. Minsky divided the issues faced in developing what he called 'heuristic' computer programming into five areas for further examination: 'search, pattern-recognition, learning, planning, and induction'.[8] His overview of the current status of AI also included a limited discussion of approaches to AI that were based on direct analogies with the brain itself. It is notable that for researchers in the burgeoning AI field, the relationship between mind and machine went both ways. In his book *The Computer and the Brain*, John von Neumann focused on the capacity of both categories of object to deal with information, to receive, process, and output data. The brain, from this influential perspective, was treated as if it were a computational organ, directly analogous to a machine.

During the 1950s and 1960s, researchers such as Minsky and Frank Rosenblatt went on to develop different methodologies for creating what they believed would, within a generation, become machine-based artificial general intelligence (AGI), with a cognitive capacity comparable to humans. This assumption proved to be over-confident. Machines have certainly outperformed humans in relation to complex tasks; they can win chess or Go and can process huge amounts of data to find patterns that are invisible to humans. Natural Language Processing (NLP) chatbots are now able to generate creative content, such as songs, stories, or sermons.[9] Yet all this success is still largely related to performance in *specific* tasks: AGI remains a theoretical topic and not a realised goal, although many researchers and commentators expect it to be attained in the near future. Other key premises of the 1950s–1960s persist in twenty-first-century debates. These include the assumption that intelligence can be measured by the ability to solve problems and that the capacity for cognition is located primarily, if not entirely, within the individual human brain. These beliefs, themselves shaped by the historical circumstances in which the idea of 'intelligence' emerged and the economic and cultural uses to which it was put, have continued to frame modern discussions of AI. They have also affected the ways in which people have anticipated and imagined the likely impact, whether positive or negative, that AI will have on human futures.

Transcendence and risk

In May 2023, the 'godfather of AI', Geoffrey Hinton, quit his job with the company Google, saying that he wanted to be able to speak freely about the dangers AI poses to human civilisation. He identified three key threats, of which the first two were bad enough: the corruption of public debate through proliferation of misinformation and social disruption via the radical reshaping of the job market. The third risk went even further. If AI proceeded to AGI, said Hinton, there would be a threat to the very survival of the human species. Hinton is not the only software professional to have raised concerns about the regulatory oversight of AI and the long-term role of AI in the human future. But what's notable in these broader debates is that concerns about the dangers posed by AI are often expressed alongside the competing claim that it is *only* through AI that humanity will be able to overcome with the multiple ecological, economic, and epidemiological threats that it currently faces. AI, so the claim goes, will 'save the world' by maximizing economic productivity, lengthening human lifespan, solving the climate crisis, curing cancer, and colonising space. The stress is on a *species-level* threat: on this reading, AI will act as the saviour of humanity as a whole.

This combination of looming existential threat with the promise of technological redemption reveals interesting assumptions on the part of AI engineers, philosophers, and commentators, some of which have a long and troubling history connected with eugenicist thinking.[10] Proponents of AI also minimise the harm that unregulated AI research is doing in the present day, whether in relation to its enormous ecological footprint or reliance on a hidden and mistreated human workforce.

Hinton's precise concern about the nature of the AI threat revolved around the fact that the kind of intelligence that is being created is collective in its very nature—and hence very different from the kind of intelligence humans have. But as we saw earlier, this claim ultimately depends on how you define intelligence and the kinds of human activity you are willing to include within that category. Many science fiction writers (John Wyndham, William Golding, Octavia Butler, the Wachowski sisters) have tried to show how different kinds of intelligence might be expressed and experienced. The 'rational' definition of the term can frequently show traces not only of species-bias, but also of its racial, gendered, and faith-based past.[11]

The philosophical movement known as transhumanism is interwoven with pro-AI philosophy, as well as sharing roots with eugenicist conceptions of intelligence. It is also a movement with many hallmarks of religion. Transhumanism was introduced to the general public by the biologist Julian Huxley in an essay written around the same time as Minsky, von Neumann, and other mathematicians were publicly debating the possible shape of AI. Huxley argued that modern science had now advanced to the point where the human species could escape its present limited and miserable condition; it could, in fact, transcend itself. By the 1980s, based on developments in genetic and computer science, some writers predicted that individuals would be empowered to use technological innovation to eliminate disease and disability, while poverty or malnutrition could be eradicated in society as a whole. This focus on overcoming the limitations of the body was only the first step in the transhuman future. For transhumanists like Nick Bostrom, Max More, and others, it is the synergies between computer science, nanotechnology, and neuroscience that will radically reshape the human condition.[12]

In 1988, for example, the computer scientist Hans Moravec published the book *Mind Children: The Future of Robot and Human Intelligence.* Moravec took von Neumann's computation model of the brain one step further, suggesting that mind–machine transfer would ultimately be both possible and desirable. This is in many ways the logical outcome of the elision between concepts such as 'intelligence', 'mind', and the individual soul discussed earlier in this chapter. Moravec's position was that the link between identity and body was illusory, since the true nature of individual identity rested in patterns which could be reconstructed and reproduced—even, perhaps, redeemed? Other scientists and futurists, including Ray Kurtzweil and Randall Koene, picked up on the theme of whole-brain emulation, whereby an individual's consciousness might be simulated through computer software and through which humanity could achieve digital immortality.[13] The economist Robin Hanson explored a slightly different perspective in his 2016 book, *The Age of Em*. Here, he considered what might happen to the structure of social life after researchers have succeeded in scanning the human brain in sufficient detail to enable 'ems' or 'emulated people' to exist in digital format.

Hanson's book suggests that some 'ems' will want their digital existence enhanced by access to robotic bodies. But for a significant number of transhumanists, the point of uploading the mind is to transcend the human condition by leaving the body behind. There are obvious parallels here, not just with the mind/body duality that is deeply rooted in Western culture, but with the specific relationship between the body and soul in some traditions of Christian belief, where the soul survives the moment of death and the immediate disintegration of the material body.

Other parallels between transhumanism and religion have been traced out by scholars such as Robert Geraci, whose analysis of what he calls 'Apocalyptic AI' examined the correspondences between the apocalyptic traditions of Judeo-Christianity and the visions of the human future mapped out by Moravec, Kurzweil, and others. Geraci's work points to the fact that ancient Jewish and Christian apocalypses both anticipated that, in his 'final intervention in history, God will overthrow the oppressors, create a perfect new world, and resurrect the righteous in purified and glorified new bodies.'[14]

The anthropologist Beth Singler has analysed other ways in which the language and forecasts used by AI thinkers draw on religious metaphors and narratives. She points to the similarities between Pascal's wager and a famous thought experiment known as Roko's Basilisk. In Pascal's wager, a rational person believes in God since the finite costs of belief are far outweighed by the eternal losses (damnation) of unbelief. Roko's Basilisk emerged on an internet forum called LessWrong, run by AI researcher Eliezer Yudkowsky, which aimed to promote rational discussion of what friendly AI might do for humanity. In 2010, a poster called Roko wondered whether an otherwise-benevolent, future AI might punish anyone who had not helped in its creation, tormenting them in virtual reality simulations. In other words, anyone who was capable of creating independent AI but did not 'believe' in it sufficiently to actually do it would face digital damnation. As Roko went on to point out, anyone who read the post and did not commit to working towards achieving AI superintelligence would certainly be subject to that punishment. Yudkowsky banned further discussion, on the grounds that the subject was harmful. As Singler pointed out, one way of avoiding future punishment would be to donate money to Yudkowsky's Machine Intelligence Research Institute, thus helping to bring about AGI. It sounded a lot like the indulgences that could historically be bought from some Catholic priests.[15]

Notwithstanding the fears and furore around Roko's Basilisk, for many transhumanists, AGI will soon empower us to create an Earthly paradise. In the longer term, we will join our 'mind children' in immortal machine bodies, free to draw on limitless computational power as we move into our new lives. Scholars such as Geraci and Singler are not arguing that the philosophers, scientists, and gamers that they study are consciously thinking in religious terms. Instead, they show that, even where religion is consciously disavowed, religious metaphors and tropes implicitly frame their understanding of the future, framing and limiting the ways in which we are all able to think about it.[16]

Technologies of salvation

In 2017, the Protestant Church in Hess and Nassau decided to celebrate the 500th anniversary of the Protestant Reformation by providing the town of Wittenberg with a robot priest. 'BlessU-2' could vary its voice to suit the priestly gender preferences of worshippers; it could give blessings in five languages and beam light from its hands. A year previously, a Buddhist temple outside Beijing had offered visitors the opportunity to interact with a robot monk, named Xian'er, created explicitly to represent 'a reflection of an innovative Buddhist spirit'.[17] There are also plenty of fictional examples of robots and other AIs creating religions, from Isaac Asimov's robot messiah Cutie to the Church of Robotology in the animated comedy *Futurama*. The natural language chatbot ChatGPT has not only provided pastors with help in developing their sermons, but has designed and delivered its own service.[18]

The prospect of becoming closer to the divine by merging with intelligent machines takes things a step further. So too does the possible emergence of religious sects based on the worship of AI, itself premised upon the potential for AI to surpass the limits of the human and to become as God in its capacity to act in the world. While there are many examples in science fiction of alien gods, these stories rarely end well.[19] What's fascinating about them in the context of AI debate is that these stories also usually share an assumption with many transhumanist thinkers that divinity derives from the combination of extraordinary levels of intelligence and power. This tends to result in science fictional situations where absolute power corrupts absolutely, usually with chaotic and apocalyptic consequences. Some writers draw parallels between AGI and the imperial history of Euro-American nations and cultures. In other words, AGI is the coming imperial power that will destroy the unprepared or the unworthy. This was the view of Victorians bringing 'civilisation' to other parts of the world, in the science-religion truce explored in Chapters 3 and 8. Syed Mustafa Ali has moreover argued that the transhumanist language of Kurzweil and other writers does not only resemble apocalyptic narratives but also bears strong similarities with the tradition of Western-Christian salvation stories. This, for Ali, links them with other historical examples of white Christian attempts to save the world, notably, the Crusades.[20]

The transhumanist focus on the human mind gains significance in the context of AGI as the saviour of humanity. A focus on the mind and its data marginalises the body and physical markers of gender, race, and class. It is not an accident that most of the people involved in discussions of future impact of AI tend to be white, male, and upper or upper-middle class. Being defined by one's mind is a privilege that extends only to those whose bodily needs are so well-supplied that they can ignore them. In other words, making everything about the brain or the mind makes it easier for discussions of AI to assume that the Euro-American experience represents a human universal. Embodied experiences of poverty, gender, and so on are supposedly irrelevant to success in this world or the future world of AGI. An example of this philosophy is found in recent projects to supply children in poor countries with laptops; access to a computer was supposed to shortcut their need for teachers or any other social infrastructure.[21] It is a small step for tech enthusiasts on the far right to blame any failure to adapt to the new world of AI on innate mental inferiority. It looks very much as though the Christian cultural context for much AI research has bred a sort of scientific theology of damnation for the non-elect.[22]

Human obsolescence

There are, as Geoffrey Hinton pointed out, less existential fears raised by the prospect of AI. In the West, the history of machine intelligence is inextricably tied to fears of social and economic disruption. From the Luddite riots that greeted the invention of the binary-coded Jacquard loom in the early nineteenth century to the impact of ChatGPT on the creative industries in the 2020s, the dread of becoming obsolete is a key theme. But as sociologists and economists have repeatedly demonstrated, automation is as problematic as AI when it comes to human autonomy.[23] By de-skilling trades and professions, automation makes it possible for managers to treat human beings as exchangeable parts within the company machine. In this way, machines not only replace human workers but also model desirable behaviour for those that remain. Western workers increasingly experience management by measurement. Surveillance through cameras and keyboards enables the analysis of performance at the granular level of keystrokes and eye movement, with profound consequences for human health and wealth. It is not coincidental that participants in the gig economy formed a key demographic in Singler's study of the social media users who claimed, and hoped, to be 'blessed by the algorithm'.[24] Whatever AI may bring for the future, those whose prayers to the algorithm are not answered can face redundance, obsolescence, and alienation in their present-day lives.

Many transhumanist writers look towards an optimistic future. These include the Order of Cosmic Engineers and the Turing Church, which anticipates the deification of AI on the understanding that it is through technology that we will find or create God. However, their hopes of transcending humanity raise the possibility of becoming alienated from humanity as it currently exists. The science fiction author Ted Chiang explored this theme in his essay 'Catching crumbs from the table', published in the journal *Nature* in 2000. Chiang describes a future in which a new form of 'metahumanity' has emerged. These beings communicate with each other via digital neural transfer, which has transformed scientific research to the point where ordinary humans can no longer understand it. Instead of original research, scientific journals like *Nature* now find themselves in the position of publishing papers on 'metahuman hermeneutics', where human scientists devote their time to trying to translate metahuman research, or to 'reverse-engineer' metahuman technology. Chiang wonders whether such activities are a waste of time, comparing them to 'a Native American research effort into bronze smelting when steel tools of European manufacture are readily available'.[25] He concludes that it *would* probably be worthwhile, since the work might lead to other intelligence-enhancing procedures that could allow humans to compete with 'metas'.

Chiang's essay, written in the first-person plural, reflects the difficult challenge of choosing which 'us' in the future is imaginatively continuous with the present 'us'. Are 'we' the legacy, biological humans of the future, who look like primitive indigenes in comparison with metahumans? Or are 'we' the metas who have successfully engaged with AGI? A commitment to the body as site of the self would suggest the former; a belief in the power of intelligence to carry 'us' through would suggest the latter. The Western notion of selfhood as mental ('I think therefore I am') suggests that the latter is likely to win out: a continuity in the ability to imagine indicates a continuity of selfhood. If we can imagine a different self, it is not different after all. The metas can be 'us'.

Although the 'we' of Chiang's essay is apparently identified with legacy humans, his final sentence constructs a bridge of identity between present-day humans and future metas just as Cartesian philosophy would suggest: 'We need not be intimidated by the accomplishments of metahuman science. We should always remember that the technologies that made metahumans possible were originally invented by humans'. In this sense, it is the metas who are the future 'us'. This, in turn, creates a sub-human 'them' in the present or in the recent past: the 'Native Americans' who don't engage intelligently with technology. It is intolerable to think of 'us' as imaginatively continuous with colonised indigenes since this would be tantamount to admitting that we are not worthy of survival or salvation. Thus, Chiang contextualises the transhumanist future of science within the history and theology of European imperialism, adopting intelligence (or the lack of it) as an agent of salvation or damnation.

The later twentieth century saw sustained challenges to the racist valorisation of 'intelligence' in the form of civil rights, equal rights, and other liberation movements. Early in 2025, the roll-back of Diversity, Equity, and Inclusion initiatives by the second Trump government and its tech-billionaire friends suggested that the underlying philosophy of white elites had by no means been undone. The possibility of maleficent AGI is, for some, a reason for uniting humanity against a common threat. Transhumanists and technofascists, however, have used the self-same entity (equally powerful, but probably beneficent) to double down on a partitioning of humanity into the saved and the damned, judged according to the quality (intelligence) of which it is supposedly a transcendent expression. Their vision is profoundly religious, but not in a good way, being indebted to colonial articulations of Christianity.

Beyond the brain

If bad theology lies at the root of unhelpful and unjust AI, then better theology is required to fix it. Science and religion can work together to create better kinds of futures. Recent theologies and psychologies of embodiment are combatting the dualist narratives of body vs soul, material vs data, which characterise common perceptions of AI. Research into embodied cognition is growing, paralleled by theological explorations of prayer, meditation, and worship as embodied and interpersonal activities.[26]

Some models of AI are called Natural Language Processing (NLP). The label suggests that machines can engage in language as an abstract enterprise, making meaning from disembodied semantic labels. A better theology of language is embodied; as Lakoff and Johnson put it in their groundbreaking analysis of metaphor, 'abstract concepts arise via metaphorical projections from more directly embodied concepts'.[27] Language and metaphor reflect, and in turn structure, embodied experience. Warm bodies give a 'warm welcome' in the form of a physical embrace. Time can be 'wasted' because it is a finite aspect of human existence. The language of possession narrates a gesture of taking and protecting: 'mine!' Living in similar bodies and in a shared environment—being subject to gravity, hunger, and fatigue—creates the common understanding from which metaphor emerges. Language itself, which has been taken both as a marker of human uniqueness and as an expression of the mind/body separation, is shaped by the body. When it comes to NLP, we would be wise to consider the psycho-social origins and the theological implications of the words (waste, possession) that are being baked into its framework of agency. Non-Western languages, with their radically different frameworks of thinking about the world (particularly in regard to ownership), offer alternative and perhaps better bases for machine governance.

Other inspirations for embodied cognition have been drawn from the philosophy of phenomenology, which investigates the structure of our lived experiences.[28] The scholars Heidegger, Husserl, and Merleau-Ponty all, in different ways, stress the physicality that contextualises our thoughts. Merleau-Ponty, drawing on Heidegger's 'being-in-the-world', goes so far as to suggest that our bodies constitute our consciousness and that the distinction between the mind and the body (a key influence in the history of Christianity as well as AI) is an illusion. Our bodies do not merely enable the input of sensory data, ready to be processed by the brain as the executor of human agency. From the perspective of embodied cognition, the concepts that shape our understanding of the world are themselves shaped by our bodies, to such an extent that we would see the world differently if our bodies were shaped differently.

A number of research programmes studying embodied cognition have sprung up. Variously described as extended, enactive, or embedded cognition, they look at the ways in which apparently abstract concepts are embodied. Some studies have demonstrated that it is easier to do tasks that are appropriately physically oriented. Thus, people react quicker in psychological tests involving object orientation if the position of the response button matches the object. In other words, they are quicker to react if the button to indicate left orientation is on their own left side, slower if it is on their right. Neuroimaging of the brain has shown mirror neurons responding when we see other people doing things, while other studies have suggested that areas associated with particular actions (kicking, running) activate when those words are read, indicating a direct connection between symbol and action. In each case, the need for symbolic cogitation seems limited. Commercially successful ventures into consumer robotics are based on similar patterns of sense/response. The Roomba robot vacuum cleaner, for example, is based on what the roboticist Rodney Brooks called the three principles of 'avoid', 'wander', and 'explore'. These are the mechanisms for generating behaviour that produce what looks like goal-directed action (moving around the floor of a room without getting stuck) and goal achievement (cleaning the floor), without any need for symbolic representation.[29]

These projects reverse the traditional orientation of AI work. Rather than starting from the top by trying to replicate or model complex cognition, researchers at MIT have instead begun with simple devices that can 'learn', in the sense that behaviour emerges out of basic principles that lead to self-organisation. As such, these devices need to interact with the world. Physical robots can move about within their environment or point to things around them, without running software. Other ventures into robotics have implications for understanding not only kinaesthetic cognition, but also the role of emotion, affect, and social interaction in human and artificial intelligence. Rosemary Picard's Affective Computing Lab, also at MIT, has been working on developing robots that can identify and respond to the emotions of the people around them.[30] This reflects key developments in the understanding of problem-solving, information-processing, and decision-making in the later twentieth century, particularly with reference to the importance of play and the significance of emotion.

The role of social interaction in the context of intelligence is perhaps even more important. Ted Chiang's fictional humans ('us') have the option of augmenting their children so that they grow up as metahuman, but only if they are brought up within metahuman culture. Faced with estrangement from their child, few human parents take this option. Chiang's stress on the importance of socialisation echoes studies in comparative psychology, which since the 1950s have shown that many behaviours once considered instinctive need to be learned through example and experience. Intelligence is not the prerogative of an individual brain, but something that is developed through interaction with others and with the environment. Social interaction shapes intelligence by providing access to the cumulative and collective knowledge of humanity.

What, then, would a social AI look like? Mark Lee opened his 2020 review of 'human friendly' AI with an account of the miniature robot Keepon. Keepon's body is made up of two yellow balls, one for the head and one for the body. It is only twelve centimetres tall and has two very small eyes (cameras) and a little nose (a microphone). It can wobble and nod its head, bounce up and down, and rock its body back and forth. This is the sum and total of its capacity to move, but humans find it fascinating. As Lee points out, Keepon is captivating 'because it *looks at humans when they talk* and ... it looks at the same objects that the human looks at'.[31] Like dogs and other companion animals, the robot gives the impression of wanting to interact with humans, a desire that is physically expressed through bodily orientation. Keepon does not have any awareness of being social, but *is* social AI in the sense that humans are prepared to treat it as if it did. Its success in charming its audiences depends on their capacity to anthropomorphise, their willingness to acknowledge the agency of others. Social intelligence is perhaps distributed across the relationship, rather than residing in one participant.[32] As such, it would be amenable to theological critique and construction.

Relational intelligence, relational theology

The modern version of the Turing test relies on the capacity of a computer to relate to humans via language, in the way that other humans would. Christian theology suggests that embodiment is at least as important as language for defining humanity, a theology that finds its ultimate expression in God's incarnation as Jesus Christ. Christian theology is also relational, again exemplified in a Christology treating Jesus as a source of example rather than rule-giving. Even Jesus' teaching, in the form of parable, can be considered as embodied and relational. As Victoria Lorrimar puts it, 'we understand through story, and narratives depend on embodied cognition.'[33] While AIs are very good at playing games, and even at generating stories, the extent to which they can learn from them is less clear. As Beth Singler notes, it is interesting to consider how AI might perform in the context of role-playing games, which demand not just that AI pass for human, but that it passes for a human who is pretending to be someone else. Could this be the twenty-first-century model for the Turing test?

This chapter's account of how AI has been conceived and developed over the past seventy years is necessarily limited. What it should have made clear, however, is that AI concepts, debates, and practices fundamentally challenge the idea that science and religion are necessarily oppositional forces. Consciously or not, AI researchers draw on theological frameworks and concepts to make sense of their work and its impact on the future. Their ways of talking and thinking have demonstrable consequences in the real world, contributing to the marginalisation of individuals and communities that are disadvantaged both historically and in the present day. The fact that religious narratives permeate these debates so thoroughly suggests an opportunity and value for doing theology. It would be salutary to take this language seriously and to see whether the tools and techniques developed by theologians would help identify and unpick some of its more problematic assumptions.[34] In tandem with this effort, theologians can learn from the widening field of embodied and relational intelligence. Knowing more about these can help to move the public discourse around AI away from the unproductive dualities of salvation/damnation and master/slave that frame it at present. Drawing on a growing understanding of human nurture and of cooperative, caring alliances with the more-than-human world may yield creative and equitable possibilities for an AI future.

Further reading

The field of AI is moving so quickly that it is difficult to give recommendations in this field. The following recommendations relate to the history of human intelligence.

Alison Bashford, *The Huxleys: An Intimate History of Evolution* (Chicago: Chicago University Press, 2022)

Alison Bashford and Philippa Levine, eds., *The Oxford Handbook of the History of Eugenics.* (New York: Oxford University Press, 2010)

Notes

1. Alan Turing, 'Computing Machinery and Intelligence', *Mind* 59.236 (1950): 433–460. Also in *Parsing the Turing Test*, edited by R. Epstein, G. Roberts, and G. Beber (Dordrecht: Springer, 2009), https://doi.org/10.1007/978-1-4020-6710-5_3.
2. Stephen Cave and Kanta Dihal, 'Hopes and Fears for Intelligent Machines in Fiction and Reality', *Nature Machine Intelligence* 1 (2019): 74–78, https://www.nature.com/articles/s42256-019-0020-9. Stephanie Dick explores the slave metaphor in early computing in *After Math (Re) Configuring Minds, Proof, and Computing in the Postwar United States* (Harvard: Harvard University Press, 2015).
3. Alan Turing, 'Computing Machinery and Intelligence', *Mind* 59.236 (1950): 433–460, p. 443.
4. George Romanes, *Mental Evolution in Man* (London: Kegan, Paul, Trench, 1888).
5. Arthur Jensen, 'Galton's Legacy to Research on Intelligence', *Journal of Biosocial Science*, 34.2 (2002): 145–172, doi:10.1017/S0021932002001451.
6. Amy Liu, 'Unravelling the Myth of Meritocracy within the Context of US Higher Education', *Higher Education* 62 (2011): 383–397, doi:10.1007/s10734-010-9394-7.
7. Charles Murray and Richard Herrenstein, *The Bell Curve: Intelligence and Class Structure in American Life* (New York: Free Press, 1994).
8. Marvin Minsky, 'Steps toward Artificial Intelligence', *Proceedings of the IRE* 49.1 (1961): 8–30, doi: 10.1109/JRPROC.1961.287775.
9. Jennifer Haase and Paul H. P. Hanel, 'Artificial Muses: Generative Artificial Intelligence Chatbots Have Risen to Human-Level Creativity', *Journal of Creativity* 33.3 (2023): 10067, https://doi.org/10.1016/j.yjoc.2023.100066.
10. Eugenics is the idea that the population can be improved by the encouragement of 'superior' people to have more children and/or the elimination of the 'inferior', usually defined by race or ill health of various kinds. Emile P. Torres, 'The Dangers of the "Galaxy Brain": A Conversation about Longtermism with Emile P. Torres', MIT Office of Sustainability, 25 October 2023, https://sustainability.mit.edu/event/dangers-galaxy-brain-conversation-about-longtermism-emile-p-torres, accessed 18 February 2025.
11. Amanda Rees, 'Rethinking Intelligence in a More-Than-Human World', *Noema Magazine*, 8 September 2022, https://www.noemamag.com/rethinking-intelligence-in-a-more-than-human-world/, accessed 18 February 2025.
12. Julian Huxley, 'Transhumanism', in *New Bottles for New Wine* (London: Chatto and Windus, 1957); Nick Bostrom, 'A History of Transhumanist Thought', *Journal of Evolution and Technology*, 14.1 (2005), unpaginated, https://ora.ox.ac.uk/objects/uuid:55ab57ec-70d0-4b93-b058-0d7f57167cc2/files/me7362f022f645636ed10948d03a7bfab, accessed 18 February 2025.
13. Ray Kurzweil, *The Age of Spiritual Machines: When Computers Exceed Human Intelligence.* (New York: Penguin Books, 2000); Randal A. Koene, 'Feasible Mind Uploading', in *Intelligence Unbound: The Future of Uploaded and Machine Minds*, edited by Russell Blackford and Damian Broderick (Oxford: Wiley-Blackwell, 2014).
14. Robert Geraci, *Apocalyptic AI: Visions of Heaven in Robotics, AI and VR* (Oxford: Oxford University Press, 2010), p. 17. For a critique of dualist interpretations of Pauline dualism, see Paula Gooder, *Body: A Biblical Spirituality for the Whole Person* (Minneapolis: Fortress Press, 2016).
15. Beth Singler, 'Roko's Basilisk or Pascal's? Thinking of Singularity Thought Experiments as Implicit Religion', *ImplicitReligion* 20.3 (2017): 279–297, doi:10.1558/imre.35900.
16. George Lakoff and Mark Johnson, *Metaphors We Live By* (Chicago: University of Chicago Press, 1980).
17. Harriet Sherwood, 'Robot Monk to Spread Buddhist Wisdom to the Digital Generation', *The Guardian*, 26 April 2016, https://www.theguardian.com/world/2016/apr/26/robot-monk-to-spread-buddhist-wisdom-to-the-digital-generation, accessed 18 February 2015; Harriet Sherwood, 'Robot Priest Unveiled in Germany', *The Guardian*, 30 May 2017, https://theguardian.com/technology/2017/may/30/robot-priest-blessu-2-

germany-reformation-exhibition, accessed 18 February 2015; see also Anna Puzio 'Robot, Let Us Pray! Can and Should Robots Have Religious Functions?', *AI and Society*, 11 December 2023, https://link.springer.com/article/10.1007/s00146-023-01812-z, accessed 18 February 2015.

18. Kirsten Grieshaber, 'Can a Chatbot Preach a Good Sermon?', Associated Press, 10 June 2023, https://apnews.com/article/germany-church-protestants-chatgpt-ai-sermon-651f21c24cfb47e3122e987a7263d348, accessed 18 February 2015.

19. The crew of the Starship Enterprise seemed to spend an inordinate of time encountering and killing beings who claimed divinity (*Star Trek V: The Final Frontier*; 'Squire of Gothos'; 'Dhardra'; 'Where No Man Has Gone Before').

20. Syed Mustafa Ali, '"White Crisis" and/as "Existential Risk", or the Entangled Apocalypticism of Artificial Intelligence', *Zygon* 54.1 (2019): 207–224, https://doi.org/10.1111/zygo.12498; Robert E. Lerner, 'Antichrists and Antichrist in Joachim of Fiore', *Speculum*, 60.3 (1985): 553–570, https://doi.org/10.2307/2848175.

21. Morgan G. Ames, *The Charisma Machine: The Life, Death, and Legacy of One Laptop Per Child* (Cambridge, MA: MIT Press, 2019).

22. Beth Singler, 'Existential Hope and Existential Despair in Apocalypticism and Transhumanism', *Zygon*, 54.1 (2019): 156–176, https://doi.org/10.1111/zygo.12494.

23. Amanda Rees (2021) 'Uncanny Valets: Machine Intelligence in East and West', *Noema Magazine*, 1 April 2021, https://www.noemamag.com/uncanny-valets/, accessed 18 February 2015.

24. Beth Singler, 'Blessed by the Algorithm: Theistic Conceptions of AI in Online Discourse', *AI and Society* 35 (2020): 945–955, https://doi.org/10.1007/s00146-020-00968-2.

25. Ted Chiang, 'Catching Crumbs from the Table', *Nature* 405 (2020): 517, https://www.nature.com/articles/35014679.

26. Tamer M. Soliman, Kathryn A. Johnson, and Hyunjin Song, 'It's Not "All in Your Head": Understanding Religion from an Embodied Cognition Perspective', *Perspectives on Psychological Science* 10.6 (2015): 852–864, https://www.jstor.org/stable/44281956.

27. George Lakoff and Mark Johnson, *Metaphors We Live By* (Chicago: University of Chicago Press, 1980), p. 497

28. Lawrence Shapiro, *Embodied Cognition*, 2nd edn. (London: Routledge, 2019).

29. Rodney Brooks, *How Robots Will Change Us* (New York: Doubleday, 2003).

30. Rosalind Picard, *Affective Computing* (Cambridge, MA: MIT Press, 2000).

31. Mark Lee, *How to Grow a Robot* (Cambridge, MA: MIT Press, 2020), p. 4.

32. For a discussion of kenotic human attribution as definitive of humanity, see the concluding chapter of Amanda Rees and Charlotte Sleigh, *Human* (London: Reaktion, 2020).

33. Victoria Lorrimar, 'Mind Uploading and Embodied Cognition: A Theological Response', *Zygon*, 54.1 (2019): 191–206, https://doi.org/10.1111/zygo.12481, p. 201.

34. The work of Margaret Masterman, who was a leading figure in both computational linguistics and on religious language, is a good example here. See Rowan Williams, *The Edge of Words: God and the Habits of Language* (London: Bloomsbury, 2014).

8

Pandemic

Metaphors matter

Metaphors matter in medical communication. Over fifty years ago, Susan
Sontag vividly analysed the ways in which the metaphors attached to certain kinds of diseases—HIV/AIDS, TB, cancer—impacted upon their public
understanding, and as a result, on the way in which people with these
diseases were treated. Since then, numerous scholars have focused on the
ways that different metaphors make disease culturally meaningful. Their
common conclusion is that the 'war' metaphor, whereby individuals and
communities must 'fight' to 'defeat' illness, is not only unhelpful but even
dangerous. Describing an illness as a battle creates an unhelpful terrain
on which to manage ill-health, especially where the path to recovery is
uncertain or unlikely: if the battle is lost, is that down to weakness—even
cowardice—in the face of the enemy? If a decision is made to pause treatment, is that a retreat? And yet, despite this scholarly consensus, military
metaphors were one of the most frequent strategies employed by politicians and journalists in the United Kingdom to communicate public health
information during the Covid-19 pandemic.

In this chapter, we examine the widespread use of conflict-based
metaphors to frame the experience of, and response to, Covid-19 in the
United Kingdom.[1] We argue that these metaphors effected a 'truce' between
science and Christianity, united against the common enemy of the virus.
UK politicians and media were able to fuse a culturally diffused version of
Christianity with Second World War (WW2) mythology in order to mobilise
and direct public trust in science, to the dismay of medics and scientists,
as well as many members of faith communities. This model of conflict, we
argue, made no more sense when wielded against a common enemy (a virus)
than it did to describe apparently opposed modes of knowledge (science and
faith). The pandemic offers a wealth of examples of how conflict metaphors
entangle narratives about science and religion, as well as the dangers of using

Science, Religion, and the Human Future. Amanda Rees et al., Oxford University Press. © Amanda Rees,
Franziska E. Kohlt, Tom McLeish, Charlotte Sleigh, David Wilkinson (2025).
DOI: 10.1093/9780191995316.003.0009

these narratives in attempts to generate behaviour considered conducive to the public good.

Military metaphors

From the earliest days of its response to the pandemic, UK government representatives 'declared war on coronavirus', framed themselves as a 'wartime government' (UK Prime Minister Boris Johnson), and described their response as a 'battle plan' (UK Health Secretary, Matt Hancock). War rhetoric emerged with renewed strength as cases rose again in September 2020, when the Health Secretary made multiple references to the 'common enemy' in a parliamentary statement. Meanwhile, a well-known poster from the First World War was in turn enlisted to carry public health messages. This repurposed the face of Lord Kitchener calling for young men to enlist as soldiers and demanding that 'Your country needs YOU'. The population was told, among other things, that their country needed them to 'keep your distance', to 'wash your hands', to 'stay at home', or 'to protect the NHS'. In some cases, Lord Kitchener himself was edited, appearing in a Union Jack face-mask. Newspapers framed their coverage around similar cultural reference points. The Prime Minister's early hope for an early reduction in infection rates was recast by the media as 'It'll be over by Christmas', a slogan made popular in the historiography of World War I. (Just as in the over-optimism of 1914, this proved not to be the case.)

The use of wartime language was not confined to elected politicians. In an episode of BBC *Question Time* on 2 April 2020, the Archbishop of York, John Sentamu, stated three times, 'We are at war', before going on to advise the Health Secretary to be 'more of a Lord Kitchener'. Meanwhile, in a rare televised address to the nation on 5 April 2020, Queen Elizabeth II made explicit reference to one of the most famous songs of WW2, 'We'll Meet Again'. The song, famously performed in WW2 by Vera Lynn, promised families and friends parted by the war that they would, eventually, be reunited with their loved ones. The song's words dominated the next day's headlines across the political spectrum of media, anchoring the coverage of public pandemic response within the context of the UK's cultural experience of warfare. The promise that 'we will meet again' was projected from Piccadilly Circus's famous billboards in London and displayed in the windows of closed businesses across the UK. The song's words succinctly captured the war-based framing for the pandemic.

And not just any war, but *the* war. In the United Kingdom, people continue to speak of 'the war' to mean WW2. This war remains a key element in British national identity, living on in myths, memorials, and rituals, and consistently invoked to demonstrate the alleged virtues of the British character. Notwithstanding the vital support and help given by the Commonwealth of Nations and other allies during the war, the myth stresses that Britain stood alone against the threat of tyranny and fascism. The nation, on this reading, pulled together as a community, displaying what was variously referred to the 'blitz spirit' (cheerfulness and making do in the face of suffering) or the 'Dunkirk spirit' (small acts of bravery that collectively saved the day). There is an established, participatory culture that accompanies this rhetoric, encapsulated in the display of key symbols and the performance of significant songs (including 'We'll Meet Again') at particular times of the year. The period preceding Remembrance Sunday, around 11 November, is particularly important, with the wearing of commemorative poppies by public figures carefully policed by the media.

Desirable behaviours in the pandemic were encouraged through the invocation of nostalgic WW2 discourse. Although there were many efforts to raise money for the National Health Service, those that gained the widest public recognition and press coverage were the ones that had the clearest military connections, such as the sponsored, round-garden perambulations of the centenarian 'Captain Tom'. An aircraft restoration company flew its Spitfire (an iconic WW2 plane) every Thursday during the early pandemic with the words 'Thank U NHS' emblazoned on its undercarriage. By the summer of 2020, people were offered the chance to have the name of a loved one inscribed on the plane, with the monies raised going to the NHS charities. Over the August bank holiday weekend of that year, which coincided with the seventy-fifth anniversary of Victory in Europe Day, the plane carried out a military-style flypast of hospitals in London and Southern England. And shortly after the Queen's broadcast to the nation, an elderly Vera Lynn re-recorded 'We'll Meet Again' as a duet with the mezzo-soprano Katherine Jenkins. Lynn and Jenkins had previously performed the song together in an explicitly military setting, the BBC's D-Day anniversary concert in 2017, together with musicians from the British Armed Forces. This recording, however, was released in support of charities working to help the UK's National Health Service.

Celebrations during bank holiday weekends of 2020 generally made connections between warfare, the pandemic, and the enduring collective spirit of the British people. The BBC orchestrated national sing-alongs for families and friends to participate in separately but simultaneously: its song lyrics page was marked by a photo montage depicting street parties with Union Jack bunting and images of Spitfire planes in the sky. On a more regular basis, householders were encouraged to stand on their front doorsteps once a week to clap and bang pots and pans in celebration of NHS workers. It was an oblique echo of the repurposing of metal work in WW2, in the melting down of garden railings for armaments.

Despite the difficulties of organising public events during lockdown, national and local government encouraged the public to create socially distanced street parties and celebrations similar to those which followed the ending of 'the war'. These events, however, often resulted in the breakdown of that social distancing and a rise in infection. 'Eat out to help out', when citizens were directed to revivify the restaurant trade, was responsible for around a sixth of the new case clusters over the summer it was in action.[2] Such consequences illustrate the risk of deploying warfare narratives in the management of a public health crisis generated by a serious, contagious disease. Government and media-supported calls for the community to come together, to meet again, or even to indulge in collective song, were the very last actions to promote public good in a situation where scientific guidance and public health messaging stressed 'isolating', 'keeping apart', and 'staying home to save lives'.

Sacrifice and service

One specific feature of wartime rhetoric is the valorisation of sacrifice. As part of daily government press briefings, on 28 April 2020, the British Health Secretary Matt Hancock spoke for the first time about the deaths of health workers, and gave clear instructions as to how their deaths should be marked:

> This morning, at eleven o'clock, we paused to remember the eighty-five NHS colleagues and nineteen social care colleagues. . . . who have lost their lives with coronavirus They are the nation's fallen heroes. And we will remember them.

The parallel with the annually repeated Remembrance Day promise to soldiers killed on the battlefield that 'we will remember them' was unmistakable. The time of eleven o'clock also echoes the timing of the annual two-minutes' silence on or near 11 November. King George VI, in his VE Day speech, had similarly woven together national concerns with military and Christian metaphors, lighting on the sacrificial dimension of their supposed congruence. The wartime king called on his subjects both to remember the 'sacrifice' that had been made to win the war, and to celebrate their freedom and independence from the 'tyranny' of a 'determined and cruel foe'.

The historical connections between the NHS and WW2 further enabled an alignment of the sacrifices made by soldiers and medics. Britain's National Health Service was founded in 1948. Not least because of its wartime birth, it was entangled with military language, expectation, and ritual. It was a response to the economist William Beveridge's call for 'an attack upon [the] five giant evils': want, ignorance, squalor, idleness, and disease. These evils arguably provided a new common enemy that enabled the continued consensus of the post-war reconstruction. As Roberta Bivins and other writers have shown, it may not be coincidental that ideas around rationing and the greater good continue to shape the expectations of staff and patients at the present day. During Covid-19, it was natural and consistent to refer to health and care workers as 'frontline' personnel; those with many years' service were dubbed 'veterans'. This framing functioned not only as an orienting narrative, but in fact directed actions, orchestrating a predictable and sometimes ritualised response that drew on established traditions of war-remembrance.

Resistance to the war narrative came from some of the very people who were being positively framed by it as heroes. In the United Kingdom, one anonymous contributor to a national newspaper said simply, 'the health service is ... not staffed by heroes ... I'm not in the army and we aren't engaged in military combat', going on to point out that they 'really don't need ... people clapping ... I don't even (whisper it) need Colonel [sic] Tom'. Another acknowledged the 'eloquent and moving' nature of military ritual, but pleaded: 'forget medals and flypasts', as the 'increasingly bombastic proposals for honouring our "sacrifice" [were] beginning to feel more burdensome than uplifting'. A group of UK medics held up a sign simply saying 'Doctors, not Martyrs' during a silent protest in front of the Prime Minister's official residence in London.

Other key workers also resisted the war narrative. Teachers, for example, were largely required by government-mandated school closures to shift to remote teaching for significant periods of the pandemic. Unlike medical staff, they could not serve on the frontlines. Press coverage from right-wing media pressured teachers to do so, even where scientific guidance and indeed government restrictions opposed and prevented it. A *Daily Mail* title page, for instance, implored 'militant unions' to 'let our teachers be heroes' and get back to the classroom. The article went on: 'magnificent staff across the nation are desperate to help millions of children get back to the classroom', thereby framing as 'magnificent' only those teachers willing to return. Those preventing them were traitors, a 'militant' opposition. Working remotely, despite the significant increase in workload that it required, was treated as the lazy option. Expressing concern based on scientific guidance was positioned as the response of an enemy within. Teacher surveys reported them as feeling 'discomfort and distress about media reports that asked them to be heroes and criticized them as villains', as 'lazy', or 'scaremongers' when they raised safety concerns. They echoed the sentiments of health workers who were expected to tend the sick despite profound shortages of personal protective equipment (PPE). Notably, some borrowed the language of war to express their fear of 'going into battle without the armour'. Use of the war narrative reinforced both a sense of vulnerability and of forced compliance.

The trope of sacrifice was a specific feature of the wartime narrative. To complain was, apparently, to baulk at the sacrifice that was being demanded of one or of one's community. It left little space for individuals to engage in legitimate dissent and that failure was to create serious problems for public health and medical science in the management of the pandemic in Britain. When it became clear that some members of government and their teams had not been adhering to the standards of sacrifice that they had demanded of others, there was an outpouring of anger. The fusion of Christian-scientific virtue was a potent weapon when it backfired.

Emphasising sacrifice enabled politicians and the media to allocate clearly defined roles to different sections of the UK population: the deserving and the undeserving. Doctors and nurses, in an abstract sense at least, were worthy; owners of cafes and restaurants implicitly less so. The United Kingdom is today a predominantly secular society. But a central plank of the moral weight carried by the military narrative in the context of Covid-19 came from the fact that war narrative is itself historically steeped in religious practice and language. A residual, cultural form of Christianity continues to mark the war dead, drawing on notions of sacrifice and remembrance that are rooted in Christian ritual. This cultural Christianity continues to exert a powerful morally directive force upon its citizens, shaping the discourse of what is acceptable and unacceptable behaviour, even in novel situations such as the recent pandemic. This moral judgement brings us onto the theme of the chapter: the truce between science and Christianity in the face of a common 'enemy'.

Cultural Christianity and the Victorian truce

The *Question Time* discussion between the Health Secretary and the Archbishop of York, mentioned above, demonstrates how the general common enemy of the virus, and the particular notion of sacrifice, brought a truce between science and Christianity. Archbishop John Sentamu encouraged the politician Matt Hancock to be 'more of a Lord Kitchener', thus suggesting that this was a time for empirical action, not prayer. In return, Hancock assured the archbishop that he would 'not cease from this fight', quoting from the hymn 'Jerusalem' and implying that this was also a spiritual quest. There was something about the pandemic that suddenly made a religious discourse feel like appropriate common ground, a holy war against an evil virus. This was not the first or the only occasion on which such a truce between science and Christianity had been observed, their values grafted together in a powerful combination.

Cultural historian Pamela Gilbert has noted in the context of an earlier national medical emergency—the Victorian cholera epidemic—that the rhetoric and 'impact of medical science ... permeates the entire culture and is inseparable from the larger political and cultural history of the nation'.[3] Christian motives to heal the sick and improve the lot of the poor undoubtedly inspired a great deal of the philanthropic social improvement of the nineteenth century. The Quakers, in particular, are noted for their innovations in housing and sanitation, aimed at their employees and factory workers.

A still more striking example of science and Christianity working in tandem comes in relation to the 'civilising mission' of the nineteenth century. As we saw in Chapter 2, British scientists, collectors, soldiers, governors, engineers, entrepreneurs, and missionaries shared one another's company in Asia, Africa, and even South America. Very often a single person might wear more than one of these hats; soldiers shot rare specimens; and entrepreneurs preached the gospel. Far from home, these Britons were thrown upon one another's company, finding that, in the challenging context of an alien land, they had a good deal in common.

Take marriage as an example: atheists and missionaries could agree that modern British customs of marriage were right and proper. Savages, by contrast, might likely practise polygamy or promiscuity. The horror of unbridled sexuality was common ground for both quavering missionary and rational railway-maker. Suddenly, the atheist scientist aligned himself with the Anglican sacrament of marriage after all. Human sacrifice was another example. Whether or not God were real, all Britons could agree that this practice was simply not acceptable. (It is, of course, entirely possible that some of the horrors of 'savage' sexuality and sacrifice were nothing more than projections of a fevered British imagination.)

The British polymath John Lubbock led the way in comparing ancient tribes in Europe with modern-day 'savages'.[4] He found many striking similarities between them. 'The savage almost everywhere is a believer in witchcraft', wrote Lubbock. 'Confusing together subjective and objective relations, he is a prey to constant fears'.[5] Savages thought, for example, that illness was caused by sorcery and could be cured by magic. Such foolish thinking could also be identified in the Victorians' own deep archaeological past; prehistoric monuments in Europe began to be explained by the supposed paranoia of 'primitive man' that the sun might refrain from returning in the spring and needed to be persuaded by architecture or sacrifice.[6]

When modern-day 'savages' beheld the accomplishments of science, they were reported to explain them in terms of their magical thinking. Tales abounded of European explorers who exploited their knowledge of forthcoming eclipses to persuade credulous local people that they were themselves divine. Or, as Lubbock noted, when missionaries introduced the printing press to Fiji, it was taken to be a God. In the later twentieth century, Arthur C. Clarke continued to make the colonially-inflected claim that 'any sufficiently advanced technology is indistinguishable from magic'.

There were at least some educable colonial subjects who could be taught the scientific view that the heavens moved according to predictable cycles and were not the works or the augurs of capricious gods. Medicine could teach them that surgery and pharmaceuticals could cure their ills, not sacrifice. In more concrete form, science needed to persuade or compel these people to accept the incursions of the railway, and the extraction of local resources, all in the name of progress. Meanwhile, Christianity could help to disabuse them of the idea that there were many gods; or that the objects of nature were animate; or that they could influence the laws of nature through ritual.

Even those who did not practise Christianity themselves agreed that science was the friend of religion when it came to mission.[7] The writer Harriet Martineau, whose faith, if any, was decidedly humanistic in form, nevertheless upheld the 'civilising mission'.[8] Lubbock made a similar compromise. Lubbock was not noted for active or vociferous religious practice and explained the emergence of religion in evolutionary terms. Yet he was quite content to invoke cultural Christianity in support of his scientific mission. Lubbock asserted the 'savage' quality of contemporary 'primitives' both for their ignorance of science and their want of religion: almost all, he proclaimed, were 'without idea of deity'. The singular nature of Lubbock's deity is significant. True religion, like science, ruled out the possibility of multiple, animist gods.

The theology that facilitated this truce was that of 'Providence', the favoured Victorian euphemism for a God whom many were loath to acknowledge publicly.[9] Providence sat somewhere between the intervening God of the theists and the retiring watch-winder of the deists. It could be acknowledged in gentlemanly fashion when things went well, but did not demand the undignified behaviour of importuning it. Above all, Providence could be thanked for the rich reserves of coal that lay beneath British turf and fuelled the machinery of progress. Geologist Roderick Murchison praised the benevolent creator for the deposits that lay so conveniently near the surface of England's green and pleasant land; his sentiments were publicly applauded by no less than Bishop Wilberforce, he of the mythologised public spat with T. H. Huxley.[10] This Providence demanded gratitude in the form of human labour and diligence in exploiting its gifts, such as the efforts of the secular saints celebrated in *Lives of the Engineers* (1862). In short, the liturgy of Providence was industry, while rational Christianity and science went hand in hand when it came to the violent mission of colonialism. Confronted by common enemies, a mutually satisfactory truce emerged.

Covid-19 and the second truce

Such a truce enjoyed a renewal during the pandemic. In 2020, Prime Minister Boris Johnson (who, like Lubbock, was not noted for his piety) described the coronavirus as 'devilish' and 'evil'. It was an 'invisible mugger' and 'physical assailant', criminal and immoral. Later in the pandemic, the persistent invocation of Great Britain's 'world-beating solutions', including but not limited to the re-branding of the 'Oxford-AstraZeneca' vaccine as simply the 'Oxford vaccine', continued the same theme. They suggested a landscape of hostile international competition, at a time when global scientific collaboration and shared strategy were needed. Johnson's metaphors can be taken in conjunction with his language in the context of Brexit. Advocating Britain's withdrawal from the European Union, he repeatedly asserted the value of freedom as somehow innate to the British character. Now, in the time of Covid-19, Britain would be free of the virus. Cultural Christianity once again positioned the Providence-favoured, virtuous British public against an alien foe.

Lockdown, however, represented a contradiction: the removal of freedom in order to be free. Johnson's performative reluctance to impose behavioural restrictions was underlined by his own refusal to abide by the rules. The rhetoric of blitz-spirited community and remembrance encouraged people to believe that they were acting in broad compliance with scientific advice even as that advice was marginalised. Collective participation in rituals such as applauding the NHS every Thursday evening occurred in tandem with an increasingly flexible approach to lockdown regulations, which was further exacerbated by a shift in UK government messaging from 'Stay Home' to 'Stay Alert'.

A disproportionate number of the British casualties in this 'war' came from ethnic minorities. Black and minority ethnic (BAME) populations were early identified as groups disproportionately affected by the pandemic. Early in the first lockdown (April 2020), BAME patients accounted for thirty-four per cent of the patients admitted to UK intensive care units but only seventeen per cent of the UK population. At the 'frontlines', things were even worse: BAME workers represented twenty-one per cent of the NHS workforce but sixty-four per cent of deaths amongst this same group.[11] As historians and others have noted, there is an on-going under-representation of non-white and non-British forces in the context of military remembrance; the same was true in the quasi-military commemoration of the 'sacrifice' made in the face of Covid-19.

Moreover, in a manner analogous to the identification of 'savage beliefs' by Victorian writers, ignorant types were spotlighted by general and medical media as resisting the holy war on the coronavirus. Suspicion fell on ethnic sub-sections of the British public for vaccine refusal. The government convened panels in March 2021 to investigate why uptake was lower amongst women and BAME communities.[12] Witnesses at the two panels emphasised the point that there are multiple minority communities in the United Kingdom and that it was not helpful to lump them together.

They highlighted the 'Eurocentric' nature of the messaging from the government and explained some of the reasons why people might be uneasy about the vaccine. Direct experiences of the consequences of racist rhetoric were crucial in forming this anxiety. Collective memory contained a litany of nineteenth- and twentieth-century examples of medical experimentation on vulnerable groups and eugenic policies to contain or eliminate people of certain ethnicities. The inquiry witnesses' implicit recommendation to engage in two-way communication with the vaccine-hesitant contrasted with government recommendations published in January 2021.[13] These spoke of 'overcoming' vaccine hesitancy rather than engaging in dialogue and recommended enrolling faith and community leaders to carry the message of vaccine necessity.[14] Having churches as vaccination centres sent a particularly powerful message. The warfare footing created an urgent drive for action, in which making time or space to help people understand the underlying medical science of the pandemic, either with regard to the function of the virus or the development of effective treatments, seemed inappropriate. They were, perhaps, encouraged to regard a vaccine as 'indistinguishable from magic'.

In this context, Kitchener's appearance in the reworking of 'Your Country Needs You' was inadvertently polarising. For much of the media, and for most of the UK population, Kitchener was an unproblematic choice to voice these messages. But Kitchener is a troubling historical figure. Not only was he closely associated with the systematic and forcible internment of non-combatant populations (mostly women and children) in concentration camps during the Boer War, but he had also presided over catastrophic measles and typhoid outbreaks in those same camps. These epidemics arose from basic lack of sanitation, neglect, and overcrowding, and they caused the death of thousands of civilians. It was, at the very least, unfortunate to use Kitchener's image at a time that coincided with a renewed public attention for the 'Black Lives Matter' movement. In the worst-case scenario, holding up Kitchener as a role model risked importing the historical baggage of eugenicist discourse into present-day public health crisis, thereby directly exacerbating vaccine hesitancy amongst vulnerable groups. Malevolent agents on social media were all too willing to exploit such errors, adopting the language of eugenics and genocide to frame their Covid-denying misinformation.

Why was vaccine 'hesitancy' such a visible front on the 'fight' against the virus? Was it because of concern for equality? No doubt this was the case. But the attribution of religious-based ignorance, that is, racism cloaked in secularism, was perhaps part of the mix. If the pandemic had not been contained, might the right-wing press have advanced to racist coverage of the risk posed by the non-white non-vaccinated? It is not unknown for minority groups to be vilified for their alleged infectiousness: Mpox and HIV/AIDS are examples. At the beginning of the pandemic, persons of East Asian ancestry experienced overt racism in relation to the Chinese origin of the outbreak. Identifying an unvaccinated sub-population in the context of a wartime narrative is an extremely risky move, potentially creating an enemy within.

To put it another way: What perspectives and priorities were lost due to focus on the non-vaccination of ethnic minorities? Take-up rates of the vaccine *did* correlate with various intersectional identities. However, it is not at all certain that these variations can be attributed as a major driver of general infection rates or identified as the over-riding cause for higher-than-necessary deaths in the population as a whole. Several decisions of Johnson's government seem to hold greater culpability in this regard. Moreover, attributing blame on ethnic grounds obscured the uneven risks that were experienced along exactly the same sociological lines (again, something that the inquiry's witnesses pointed out). Immigrants and people from ethnic minorities were over-represented in the health-care professions, amongst so-called key workers (transport, for example) and in the hospitality industry that was put so recklessly at risk via 'eat out to help out'.

One final similarity between the two truces concerned injunctions to use 'common sense' in defeating the virus. This too played into Victorian tropes. In Lubbock's view, and the view of many others, 'savages' lacked common sense to understand their world. It was common sense to build a railway, to plant bigger fields, or to worship a single God. Common sense, as has often been observed, is anything but universal or rational. In the case of Covid-19, it was revived within the United Kingdom as a particular British stereotype of rationality, similar to the colonial common sense that saw aspirin as better than witch doctors.

Establishment Church leaders called out the inappropriate personification of Covid on the BBC's regular early morning religious broadcast *Thought for the Day*. They pointed out the potential for a war metaphor for the pandemic to 'other' particular people or groups in just such a way as vaccine hesitancy might have done. However, when Anglican bishops were asked to reflect on the theology of the pandemic, they were univocal in singling out an unexpected enemy: people who explained the pandemic as an act of God.[15] Here, again, was an eagerness to bruit the truce with science: God had no part in understanding the situation. It is not clear from the interviews whether this belief was really something that the bishops had encountered much. Bishops spend a limited amount of time with ordinary churchgoers, still less so in a period of social isolation. Research into religious responses to Covid-19 consistently showed a preference for such narratives only in highly orthodox conservative religious groups, whether based in Islam, Judaism, or Evangelical Christianity. In the latter case, this preference was present in circles close to former US president Donald Trump, but in each case, the 'Others' whose divine punishment had caused the pandemic were relatively interchangeable—pride parades or Black Lives Matter protests. By closing ranks behind 'science', the Anglican bishops missed an opportunity to give a strong message about the role of extractivist capitalism in the genesis of zoonotic diseases, amongst other features of the ecological crisis.

As we noted at the outset of this chapter, the use of warfare narratives in medicine is by no means confined to the Covid-19 pandemic. Evidence that their use in public and private health settings is deleterious has emerged previously in studies of HIV/AIDS, cancer, Zika virus, and a broad variety of chronic health discourses. Researchers have shown that war metaphors create an artificial win-lose dichotomy' and that, by obligating a fight, we may encourage the pursuit of futile or harmful options, stigmatising hesitancy or periods of contemplation on the part of the person or community with the disease. War-based narratives transpose 'military virtues', such as courage and perseverance, into contexts in which they have no currency and suggest the existence of options which medical reality does not recognise. Virtue, whether Christian, scientific, military, or patriotic, cannot on its own alter the nature and course of a disease.

The war-based narrative for Covid-19 was moreover remarkably similar to the Victorian truce between science and Christianity, constructed in relation to a common enemy. It was expressed as a cultural Christianity of rationality and a complementary culture of science as virtue. This truce allowed the colonialism that was common to both cultures to be expended upon common enemies, distributing blame and marginalising certain groups. This chapter has emphasised human instantiations of the 'other' that fell victim to military discourse in the Covid-19 pandemic. A longer discussion would inquire into the wisdom of identifying an aspect of nature—even such a small and specific one as a particular virus—as an 'enemy'. As eco-humanities scholars argue, 'man vs nature' is neither a good way of understanding our history nor negotiating our future.

The narrative analyses of war metaphors in the Covid-19 pandemic emphasise a broader need to make the functions of narrative in public communication about science much more evident. Scientific information and advice based on scientific research are often treated as if they are objective, detached, and value-free, and thus readily applicable in any social context. In practice, the language through which they are expressed often carries within itself problematic history and assumptions. This discourse can in turn become an active agent in forming understanding and directing behaviours. Religious leaders and institutions have their part to play in highlighting where things are framed unhelpfully and in proposing how they can be made more inclusive. To facilitate this contribution, the established British Church and mainstream Protestantism have work to do in unpicking their nineteenth-century truce with science and its consequences.

Further reading

Alan Bleakley, *Thinking with Metaphors in Medicine: The State of the Art* (London: Routledge, 2017)

Pamela Gilbert, *Cholera and Nation: Doctoring the Social Body in Victorian England* (Albany, NY: State University of New York Press, 2008)

Franziska E. Kohlt, 'A "War" against a "Devilish" Virus: Religious Rhetoric and Covid-19 in the UK', in *RecordCovid19: Historicising Experiences of the Pandemic*, edited by Kristopher Lovell (Berlin and Leiden: DeGruyter, 2023), 91–108

Laura Otis, *Membranes: Metaphors of Invasion in Nineteenth-Century Literature, Science, and Politics* (Baltimore: Johns Hopkins University Press, 2000)

Charles E. Rosenberg, and Janet Lynne Golden, eds., *Framing Disease: Studies in Cultural History* (New Brunswick: Rutgers University Press, 1992)

Anita Wohlman, *Metaphor in Illness Writing: Fight and Battle Reused* (Edinburgh: Edinburgh University Press, 2022)

Notes

1. Parts of this chapter are discussed in Franziska E. Kohlt, 'A "War" against a "Devilish" Virus: Religious Rhetoric and Covid-19 in the UK', in *RecordCovid19: Historicising Experiences of the Pandemic*, edited by Kristopher Lovell (Berlin and Leiden: DeGruyter, 2023).
2. https://www.theguardian.com/business/2020/oct/30/treasury-rejects-theory-eat-out-to-help-out-caused-rise-in-covid, accessed 18 February 2025.
3. Pamela Gilbert, *Cholera and Nation: Doctoring the Social Body in Victorian England* (Albany: State University of New York Press, 2008), p. 4.
4. John Lubbock, *Prehistoric Times: As Illustrated by Ancient Remains and the Manners and Customs of Savages* (London: Williams and Norgate, 1865).
5. John Lubbock, *Prehistoric Times: As Illustrated by Ancient Remains and the Manners and Customs of Savages* (London: Williams and Norgate, 1865), p. 470.

6. Charlotte Sleigh, 'When the Sun Goes Down', *Wellcome Stories* (2022), https://wellcomecollection.org/stories/when-the-sun-goes-down, accessed 18 February 2025.
7. Sujit Sivasundaram, *Nature and the Godly Empire: Science and Evangelical Mission in the Pacific, 1795–1850* (Cambridge: Cambridge University Press, 2005), p. 12.
8. Deborah A. Logan, *Harriet Martineau, Victorian Imperialism, and the Civilizing Mission* (London: Routledge, 2016). Martineau also had ethical qualms about the treatment of those on the receiving end of the mission.
9. John Lubbock, *Prehistoric Times: As Illustrated by Ancient Remains and the Manners and Customs of Savages* (London: Williams and Norgate, 1865), pp. 489–491. The 'rational' religion born of this truce stripped out mysticism and embodied ritual, even diminishing the historically orthodox theology of trinitarianism.
10. James A. Secord, 'King of Siluria: Roderick Murchison and the Imperial Theme in Nineteenth-Century British Geology', *Victorian Studies* 25.4 (1982): 413–442; pp. 339–340. http://www.jstor.org/stable/3826980.
11. Tim Cook, Emira Kursumovic, and Simon Lennane, 'Exclusive: Deaths of NHS Staff from Covid-19 Analysed', *HSJ*, 22 April 2020, https://www.hsj.co.uk/exclusive-deaths-of-nhs-staff-from-covid-19-analysed/7027471.article, accessed 18 February 2025.
12. https://committees.parliament.uk/work/1046/take-up-of-the-covid19-vaccines-in-bame-communities-and-women/publications/, accessed 18 February 2025.
13. https://assets.publishing.service.gov.uk/media/6001808ed3bf7f33af7bdc20/s0979-factors-influencing-vaccine-uptake-minority-ethnic-groups.pdf, accessed 18 February 2025.
14. For an account of positivist invocations of science in science communication, see Jean-Baptiste Gouyon, Franziska E. Kohlt, Kristian Nielsen, Charlotte Sleigh, and Cristiano Turbil, 'Science Communication and Scientism: Historical Perspectives', in *Science Communication: Taking a Step Back to Move Forward*, edited by Martin W. Bauer and Bernard Schiele (Paris: CNRS Édition, 2023).
15. Zara Thokozani Kamwendo, 'Resistance to Narratives of the COVID-19 Pandemic as an Act of God', *Zygon* 56.4 (2021): 1110–1129, https://doi.org/10.1111/zygo.12732.

PART III

FUTURES

9

Secular saints

My first heroes

On Christmas day 2014, the astrophysicist and science communicator Neil deGrasse Tyson put out a mischievous tweet. Ostensibly wishing Sir Isaac Newton a happy birthday, he asked his followers to join him in celebrating the fact that, 'On this day long ago, a child was born who, by age thirty, would transform the world'.[1]

In drawing this analogy between Newton and Christ, deGrasse Tyson was following in a longer tradition of the physicist's transformation into a secularised messiah.[2] The conditions of Newton's birth and early life did seem to mark him out as destined for extraordinary accomplishments. Born on Christmas Day, and like heroes of antiquity as well as many saints having no earthly father (his mother was widowed during her pregnancy), the young Isaac showed remarkable precocity in his physical and mental development. According to Newton's own claims, issued to protect his intellectual priority, most of his major discoveries were made in his *annus mirabilis* of 1666, while he was still extremely young—even younger than Christ's age at entry into public ministry.

This approach to Newton is unusual only in that it makes the comparison to Christ explicit. As we have seen in previous chapters, science itself is frequently regarded as something that is different from other forms of human endeavour, a unique activity that depends for its success precisely on its capacity to transcend subjective or biased human judgement—to take the 'god's eye view'. This presumption has important implications, not least for the kind of people that are thought to be best suited to doing science. If science is an exceptional kind of human activity, then does it follow that scientists must be exceptional kinds of human beings?

If one looks at the portrayal of scientists in popular culture, then it would indeed seem that they are not like the rest of us. Just as it has not been particularly cool to be 'religious' for the past century or thereabouts, so scientists are depicted or dismissed as geeks and nerds. Books for children

Science, Religion, and the Human Future. Amanda Rees et al., Oxford University Press. © Amanda Rees, Franziska E. Kohlt, Tom McLeish, Charlotte Sleigh, David Wilkinson (2025).
DOI: 10.1093/9780191995316.003.0010

(often with titles such as *Scientists: My First Heroes* or *Scientists are Saving the World!*) try to show science as an aspirational identity for their readers, but seemingly to little avail.[3] Since the late 1950s, a substantial body of work on public (especially children's) perceptions of 'the scientist' has shown that scientists are consistently imagined as physically, intellectually, and socially distinctive.[4] They wear glasses and white coats; they have funny hair and beards; they are clever and incomprehensible; and they are isolated and secretive, surrounded by lab equipment. The American comedy series *The Big Bang Theory* (2007–2019) embodies these stereotypes, the intellectual brilliance of its scientist characters balanced by their pronounced social ineptitude, economic naivety, and difficulty in navigating relationships.[5]

These scientist stereotypes have at least two important consequences. First, if people do not recognise themselves in the portrayal of a particular community, they are less likely to think of themselves as (potentially) part of that community. Scientific organisations have recognised this problem and have made efforts to encourage people from diverse backgrounds to study science as well as helping employers to facilitate their career progression.[6] Books such as *Scientists: My First Heroes* are a part of this well-intentioned effort, often highlighting gender and ethnic diversity amongst their selected examples.

The second problem is more insidious. These stereotypes, and the efforts made to overcome them, usually focus on the achievements of exceptional individuals, treating them as the ultimate source of scientific success. This tendency, whether shown in the fictional exploits of the *Big Bang Theory*'s Sheldon Cooper, or the real-life work of Stephen Hawking (who made several cameo appearances on the show), obscures the fact that science is, in fact, a collective enterprise. Individual achievement depends on the collaborative work of colleagues, a category that includes administrators and cleaners, as well as technicians and students—not to mention domestic support from families and friends.

As this chapter will show, this emphasis on brilliant, often unconventional, individuals as the key drivers of scientific progress is rooted in an undeclared relationship between science and religion. Earlier, we discussed how problematic the unguarded use of theological metaphor and narratives could be when used in debates about artificial intelligence. Here, we will examine the consequences of using theological language to describe scientists, and of the tendency to treat scientific biography as a kind of secular hagiography, with clear parallels in writings about the lives of the saints. We will look at the examples of Isaac Newton and Stephen Hawking, among others, to see how the idea of a scientific genius developed out of the concept of the saint, and explore the functions that their life-stories are meant to fulfil. We make the argument that secular hagiographies, rooted in the historic relationship between science and religion, shape not only how we think about scientists but also how we think about the nature of science itself. Stories of brilliant and unconventional individuals have created a story about 'the scientific method' and the grand narrative of scientific progress which do science a disservice.

Let us now praise famous men

The historian Anna Maerker has researched the writing of stories about 'great men of science'. She identifies a clear hagiographic influence in biographies from a very early stage in the history of science.[7] Maerker argues that invoking religious authority helped the writers of biographies in their underlying ambition, to create a foundation story for their particular scientific disciplines. Nineteenth-century historians of science, often themselves practitioners of science, took Thomas Carlyle's 'great man' approach as a model, honouring the achievements of their predecessors and presenting them as models for aspirational future generations.[8] This aspect of scientific biography was, of course, not confined to the nineteenth century; as we have already noted, biographies of scientists of colour, scientists with disabilities, female scientists, and those who belong to the LGBTQ community have all been used more recently to encourage increased diversity in science.

But there is an important distinction to be drawn here—a distinction to which we will return later in the chapter—between identifying someone as worthy of recognition because of their intrinsic qualities and honouring someone because of their capacity for hard work and the personal sacrifices they have made in disciplining themselves to attain a particular goal. Nineteenth- and twentieth-century heroes of science were respected and venerated because commentators believed that their work had significantly improved the human condition; their sacrifices for the good of humanity deserved acknowledgement. At the same time, even as the idea of an over-arching service to humanity structured many biographies, the lives of individual scientists provided disciplines with origin stories that could potentially be appropriated, and therefore supported, by national mythmaking.

Nicolaus Copernicus, who was born in the city of Toruń, is a good example of this. Today, Toruń is recognised as one of the oldest cities in Poland. Historically, however, it has been part of both Poland and Prussia, and over time it has seen significant immigration of German-speaking people. Shortly before Copernicus' birth (to German-speaking parents), the city was the site of active military conflict between the Prussian Confederation and the Kingdom of Poland: later on (in 1793 and again from 1815 until the end of World War 1), Toruń was annexed by Prussia. Copernicus himself went on to spend most of his adult life working in Ermland, then a province of East Prussia. As a result, the nationality of Copernicus has been hotly disputed between Germany and Poland, most notably during the Second World War, when in the teeth of his annexation by Nazi Germany, he was re-claimed by Poland.[9] Anniversaries of births, deaths, publications, and observations can become the site of confrontations and warring narratives as different communities seek to stake their claim to particular scientists. In the process, these individuals are set apart from the common run of humankind, as heroes, as saints, and—in the years following the death of Isaac Newton—as geniuses.

Newton has been celebrated both as the first scientist and latterly (in the aftermath of John Maynard Keynes' purchase of Newton's Portsmouth Papers in 1936) as the 'last of the magicians'. But his legacy also embodies a significant change in the English-language interpretation of the term 'genius'.[10] Originally referring to the tutelary or prevailing spirit of a person or place (*genius loci*), by the early modern period, 'genius' had come to be used in reference to an individual's particular (ingenious) talent or skill. As the historians Patricia Fara and Richard Yeo have shown, the modern understanding of the term, referencing an individual of exceptional creativity and brilliance, emerged as a result of the work done by eighteenth- and nineteenth-century writers on Newton's life and legacy.[11] Prior to this shift, as Robert Iliffe has argued, the most appropriate language available to discuss Newton's achievements was theological in origin. As a result, the near-divine nature of his accomplishments was repeatedly emphasised.

As Iliffe notes, Newton's followers treated him as a particularly Protestant kind of saint.[12] While he left no corporeal relics (other than his hair) and would have been utterly appalled at any kind of idolatrous reverencing of his remains, influential commentators such as Edmond Halley and John Conduitt drew directly on hagiographic techniques and values to describe and celebrate his life and works. As we saw earlier, it was possible to draw clear parallels between the early lives of Newton and Christ.

Following Newton's death, praise and admiration for his intellectual accomplishments was thoroughly intertwined with tributes to the excellence of his moral character. For example, Conduitt, the husband of Newton's niece and the custodian of his estate, offered Bernard de Fontenelle an account of Newton's life to be read to the French Académie. Iliffe describes how Fontenelle's eventual text:

> laid out the extraordinary virtues of the pious natural philosopher and lauded Newton's humility, modesty, honesty, generosity, temperance, affability, mercy to man and beast, patience and perseverance—most of which must have provoked great mirth (or incomprehension) among Newton's enemies... 'Posterity,' he continued 'will hardly believe so many virtues & no vices could exist in any man.'[13]

In 1806, Edmund Turnour published Conduitt's memoir, adding to it some words from Newton himself. This quotation, emphasising the philosopher's apparently self-deprecating humility, was to become famous:

> I do not know what I may appear to the world; but to myself I seem to have been only like a little boy, playing on the sea-shore, and diverting myself, in now and then finding a smoother pebble or a prettier shell than ordinary, whilst the great ocean of truth lay all undiscovered before me.[14]

This modest claim might have come as something of a shock to those—John Flamsteed, Robert Hooke, Gottfried Leibniz—who had been engaged in prolonged and bitter contention with the man.

The rediscovery of Newton's rivalry with Flamsteed, along with the suggestion that Newton might have been suffering from mental illness, prompted a vigorous renewal of interest in Newtonian relics in the early nineteenth century, alongside an enthusiastic defence of both his intellect and his virtue. In 1837, for example, the astronomer Charles Turnor began the *Collectanea Newtoniana*, which is now in the archives of the Royal Society and includes pictures and descriptions of Newton himself along with objects and places associated with him. Other Society members presented it with mementos such as busts, death masks—even sundials believed to have been made by Newton. Bodily relics of Newton were also on the move, in the form of locks of his hair taken from his body when it was translated to a new resting place in Westminster Abbey. Many years later, in 1979, these remains were subjected to chemical tests to see whether heavy metal poisoning could explain the symptoms of depression and irrationality that Newton seemed to have shown in his fifties.[15]

The nineteenth century was also the time in which the philosopher and social theorist Henri de Saint Simon proposed that founding a new, secular religion based around Newton would lead to greater equality, stability, and efficiency in society. In his *Lettres d'un habitant de Genève* (1802), he suggested that a council of experts drawn from the scientific and philosophical world should meet regularly (and preferably at Newton's tomb) to determine the spiritual and technical governance of society. If not at this stage a saint, Newton had become a tutelary spirit, a source of guidance in the support of rationality and reason.

Newton's contemporaries and successors understood his achievements in terms of the divine. The topics of his study aided the connection; he had been able to identify, chart, and understand the laws of nature laid down by God. He could explain how the heavens worked. As Newton's tomb in Westminster Abbey proclaimed, he had 'a strength of mind almost divine'. The language of sainthood was used to describe and extol Newton's work, with the result that theological implications crept into the modern concept of genius as it developed in the following years. With these theological implications came tensions and questions. Was genius a divine gift, as the Romantics might suggest? Or was it something that had to be enacted via effort? The question resembled the tension between election and effort in Protestant theology, with those who believed salvation was predestined at birth in conflict with those who believed that some human action was required to invoke it. Genius, then, might be read in two different ways, just like salvation. It could be seen as a spur to effort, or as the manifestation of divine favour that cannot be claimed or controlled. This is exactly the unresolved conundrum of *Scientists: My First Heroes* and their ilk.

A dialogue between an adult and a child in one of the didactic tracts that was produced by the Society for the Diffusion of Useful Knowledge in the 1820s perfectly encapsulates this tension. In *Buds of Genius*, a child named Henry asks his mother whether Newton was a good man as well as a great philosopher. His mother offers, as evidence of Newton's good character, his great modesty in claiming that his discoveries were down to hard work and patience. Young Henry vehemently disagrees:

> But I think, in that opinion he was much mistaken. If I were to be take as much pains as Sir Isaac Newton did, I do not think I should be able to make any discovery at all. You know, Mamma, he was a *great genius*.[16]

Even if Newton had conducted himself admirably, it was increasingly felt by the early nineteenth century that his genius was born, not made.

Theologies of genius

Newton was not the only philosopher whose individual character was considered relevant in assessing his intellectual contributions. Descartes was supposed to have shown the patience of a saint when dealing with his critics. (In another instance of scientific nationalism, however, Thomas Chalmers claimed in 1815 that Newton's greater success lay in the fact that he was humbler than Descartes.)[17] The connection between a person's intrinsic moral character, their innate brilliance, and their scientific achievements has persisted since Newton's death and has been interpreted in very different ways over that period.

The chemist and theologian Joseph Priestley strongly dissented from the kind of views expressed by the child Henry, outlined above. He argued for an egalitarian approach to knowledge: Newton's *Principia* might be extremely hard for most people to understand, but was this a failing in the audience or its author? If Newton had given a fuller account of how his discoveries were made, including its mistakes and accidents; if he had followed the example of Francis Bacon, a key exponent of the empirical method; would his audience have grasped his ideas more clearly? As Priestley put it,

> ... if, when a man publishes discoveries, he, either through design, or through habit, omit the intermediate steps by which he himself arrived at them; it is no wonder that his speculations confound others, and that the generality of mankind stand amazed at his reach of thought.[18]

Newton, from Priestley's perspective, had been guilty of this sin. Many, if not most, people would find advanced mathematics hard to understand, but a Baconian philosophy of science based on disciplined *doing* could provide a model for all to follow. In the later nineteenth century, a successful series of biographical portraits, *Lives of the Engineers*, was based on a similar philosophy. Following on from the even more successful *Self-Help* (published in 1859, it famously outsold Darwin's *Origin*), it recommended habits of mind and action that would enable readers to emulate the achievements of such great men as George Stephenson and James Watt. This was hagiography as self-improvement.

As the idea of the born genius took hold, however, Priestley's view fell out of favour, at least for scientists if not for engineers. Questions were raised as to whether discipline could account for all scientific achievements. Newton seemed to be an example of the kind of mind that could make original and inspired intellectual leaps that were not subject to rules of method or discipline. As one of Newton's nineteenth-century biographers suggested, scientific genius seemed more akin in its creativity to poetic or artistic genius. And, as the dissipated 'geniuses' Percy Shelley and Lord Byron came to illustrate, this kind of genius almost seemed to rule out the need for commendable behaviour. Such dull duty was for ordinary folk; genius was decoupled from virtue.

There have been a handful of flamboyant scientists who have made a bid to be considered the Byron of science. Humphry Davy is probably the best candidate of the nineteenth century, well-known for his dazzling, risqué lectures. He even wrote poetry. But in general, the decoupling of personal morality from genius worked out differently in science than it did in the arts. For one thing, it occurred in a different context. While poets and artists could more or less continue to function on their own in drawing rooms or garrets, science became professionalised and institutionalised, in shared spaces of training and research. Within these spaces, practitioners were required to maintain objectivity and disinterestedness in relation to their subjects. The greatest virtue of all was to disavow all moral and emotional connection to the work that one was performing.

This depersonalisation of research would become as important to the maintenance of public scientific authority as the display of the admirable achievements of Newton, Galileo, and others had been in earlier centuries. The idea of genius was now increasingly linked with an otherworldly eccentricity, a lack of connection with ordinary life that manifested itself in an unconventionality that was sometimes deliberate, sometimes absent-mindedly expressed. Thus, the physicist Gerald Holton described the 'chosen ones' of science as having a 'legendary' ability to focus in on a problem for extraordinary lengths of time, coupled with an 'uncompromising rejection of every ... external, arbitrary authority, in physics as well as in clothing or in the demands of everyday life'.[19] The new saints of science were usually solitary men, their great talents often existing at the expense of their abilities in ordinary things—in particular, the capacity to communicate with others. Albert Einstein, writing in honour of his fellow physicist Max Planck, described him as one of the rare inhabitants of the 'Temple of science' that is guarded by the 'angel of the Lord'. Such men, said Einstein, are 'odd, uncommunicative solitary fellows', who like painters and poets are driven to realise an understanding of the world around them.[20]

Geniuses, much like the saints, possessed the ability to push themselves beyond the limits of the body, showing what might seem a superhuman ability to ignore the need for sleep and sustenance, as well as social convention, in their pursuit of knowledge. Newton poked himself in the eye with a needle. Humphry Davy, a later leader in the field of dangerous self-experimentation, tested the new gas nitrous oxide by drinking a bottle of wine 'in eight minutes flat and then inhaling so much gas he passed out for two hours'.[21]

In other ways, the connection between scientific discoveries and the religious or moral disposition of the practitioner remained. William Whewell drew a moral distinction between first-rate researchers who could discover the laws of nature, and second-tier scientists who were only capable of derivative, deductive work. He argued that the capacity to think inductively, that it, to recognise the patterns underlying apparently disconnected evidence, emerged out of a religious disposition:

> This step so much resembles the mode in which one intelligent being understands and apprehends the conceptions of another, that we cannot be surprised if those persons in whose minds such a process has taken place, have been the most ready to acknowledge the existence and operation of a superintending intelligence, whose ordinances it was their employment to study.[22]

Individuals who are capable of doing outstanding intellectual work, according to Whewell, 'have been peculiarly in the habit of considering the world as the work of God'—that is, as fundamentally meaningful and apprehensible.[23] Fewer scientists held such views at the end of the nineteenth century than had done so at its beginning; nevertheless, the tension between the respective roles of industrious labour and brilliant creative insight on the other remained. Geniuses still needed to work hard; their vaunted capacity to ignore physical needs might indeed cause them to work harder than most. But was hard work alone sufficient to deliver superior insight? Did those scientific heroes who reframed the world have, as Newton's tomb put it, a 'strength of mind almost divine'?

Relics and legacies

Does the choice of metaphor in describing scientists matter? In Chapter 7, we saw that in the debates surrounding artificial intelligence, metaphors could very much have real-world consequences. But does it matter in relation to the 'saints' of science? Is the use of the language of religion and theology consequential, or is it simply one of the few registers available in which to note, even in a largely secular community, the outstanding contributions of an individual?

One reason why it matters is precisely because this practice does focus attention on the behaviour and achievements of an individual, rather than a community, an issue to which we will return in the final section of this chapter. It also matters that the veneration of individual scientists is not just expressed in words, but in the treatment of objects—specifically, their material remains. These objects bear, at least on the surface, a striking resemblance to the relics of saints and other holy individuals.

Objects linked with the life or death of notable scientific figures have consistently been singled out for care and preservation by their posterity. These objects can range from death masks and preserved body parts (fingers, brains, locks of hair, bones) through portraits, busts, and statues. They include personal papers, chalkboards, clothing, and desks: even the final breath of Thomas Edison, allegedly preserved in a test tube.[24] Such items have been sold for large sums of money at auction to private collectors and are on display at museums across the world. The museums themselves often resemble ecclesiastical buildings in their structure and display, as in the case of Oxford's Museum of Natural History, deliberately designed to reflect the relationship between science and natural theology.[25] Secular relics can descend into bathos as easily as the religious kind; visitors to the 200th-anniversary Darwin exhibition at the Natural History Museum, London, were greeted by the chair upon which the biologist sat to write his book, implicitly invited to consider via its dented velvet the presence of the bottom of genius.

In a review of the book *Savant Relics*, which explored the treatment of the physical remains of eminent scientists, Robert Hicks denied that such items have a religious overtone. He took as his case study the microscope slides of samples of Einstein's brain held by his own museum (the Mütter Museum in Philadelphia). While these had certainly featured in exhibitions 'about the brain as cultural metaphor and biological specimen', he was at pains to point out that no one called them 'relics'.[26] Ludmilla Jordanova gives a more generous and capacious definition: the term 'relic' is 'best treated as a term of thinking... about items of material culture associated with special people'.[27] Scientists' bodily parts retained initially for research purposes can transform into something else; the bladder of the eighteenth-century physiologist Larazzo Spallanzani, for example, was later put on public display in a museum alongside the preserved head of the anatomist Antonio Scarpa. Pieces of the desk 'at which Hawking contemplated the mysteries of the universe' were embedded in a limited-edition £20,000 watch, alongside a piece of meteorite and an 'etching of the stars from the night sky in Oxford on the date that Hawking was born'.[28] Galileo's middle finger went on display in a 'magnificent reliquary ... identical in shape to those in use in Catholic rites and exhibits'.[29]

While this veneration of bodily remains or belongings dates back to the heroic age of antiquity, it was Christianised in the West by the second Council of Nicaea (787). At this meeting, it was recognised that relics were at the heart of many churches and communities. No liturgy exists for the making of religious relics; they are recognised through entirely grassroots process and their capacity to represent a physical connection to the spiritual realm has ensured their survival through many church reforms.[30] Importantly, a relic is not an object of worship in and of itself. Theologically, it is the vehicle through which divinity can work to demonstrate the holiness of a saint. Sociologically, it brings a community together; understanding its display and use can illuminate the attitudes and assumptions that structure that community. Relics of the saints provoke complicated emotional and intellectual reactions from those who seek them out or are otherwise associated with them, ranging from awe and admiration to adoration, as well as evoking specific bodily and physical reactions associated with piety and devotion. While not themselves the agents behind extraordinary events, they are associated with miraculous happenings, which can be taken as evidence of God's will to demonstrate the holiness and sainthood of their owner.

While saints' relics are often involved in the making of miracles, in contrast, the relics of scientists are not normally linked with violations of natural law. Hawking's desk, Einstein's brain, an apple tree, grown from a cutting taken from Newton's tree at Woolsthorpe Manor, are all, in this sense, rational relics. Even Edison's last breath could be interpreted in that manner, although it represented an exhalation rather than inspiration. Through their display, audiences are enabled to feel a sense of connection with a historic scientist, and significantly, given our discussion of the often-disembodied concept of intelligence in Chapter 7, to feel an *embodied* sense of connection.

Despite the rise of objectivity as scientific virtue, it is through the bodily needs and weakness that humans hold in common that individuals singled out for the brilliance of their minds are venerated.

Rational relics help mediate the humanity of scientists, often obscured by the stereotypes surrounding scientific research. A curious tension can therefore be identified: on the one hand, the use of the language of sainthood and religion emphasizes the distance between brilliant scientific minds and the intellectual capacities of ordinary people, while on the other hand, the respectful exhibition of material relics stresses the body in which those minds were housed. This mind/body tension is perhaps demonstrated to best effect in the legacy of the English cosmologist and theoretical physicist, Stephen Hawking.

Embodied communities

Amongst the global response of grief and loss that marked Stephen Hawking's death in March 2018, one particular meme caused significant international outrage: the notion that death had somehow freed the scientist from the wheelchair which he had used since the 1960s. Pictures circulated on social media, some showing the abandoned empty chair underneath the night sky, others depicting Hawking getting up from the chair and stepping up to the stars. They even included images of the scientist standing beside his chair in front of the gates of heaven and arguing with St Peter. These images and metaphors of emancipation from disability—although notably, they continued to depict Hawking wearing his glasses—were rightly criticised for their ableist attitudes. But they were not merely offensive. They also profoundly misrepresented Hawking's own attitude to his lived experience.

For Hawking, his wheelchair, computer, and speech synthesiser were not just assistive technologies; they were crucial components of his personal and scientific identity. For example, in his public lectures, Hawking would consistently make self-deprecating jokes about the strange accent that his speech synthesiser gave him. He had adopted this technology, portable and attached to his chair, in 1986, after a tracheostomy permanently damaged his voice box, and by the early 2000s, the original sound card had begun to degrade. Voice technologies had moved on considerably in the intervening thirty years; yet despite the jokes about his synthesised voice, Hawking consistently rejected efforts to give him a new, more natural accent and tone. Instead, he insisted engineers write software that would recreate the original sound. That instantly recognisable voice was his, no matter how artificial or robotic it sounded in the age of Siri or Cortana.

Hawking had been the perfect incarnation of the idea of the scientist as a disembodied mind. He was often seen, as these images revealed, as a pure mind acting through technology: the computers he used for research and the interface that rendered his speech. Indeed, Hawking's chosen field of theoretical physics is commonly taken to be the acme of science both in terms of its difficulty and its purely mental quality. Colleagues and journalists portrayed Hawking as living with no bodily distractions or interruptions between his mind and the universe; like a monk who has achieved dominion over his bodily needs, he could achieve enlightenment. The incapacity attributed to Hawking's body was taken as fulfilment of the notion that bodies are not relevant to scientists: that science is not an embodied (or economic, or ethical) activity.

The freedom memes also revealed a misunderstanding of the networks and communities that made Hawking's intellectual achievements possible. For many observers, Hawking was both the ultimate example of the modern scientific genius and the embodiment of the rational, disembodied, individual mind. Once again, these tropes often structure books such as *Scientists: My First Heroes* in which a scientific disposition (or genius) enables the hero to rise above the limitations of race or gender. Like many popular accounts of science, they present success in the field as the result of individual effort. As we noted earlier, celebrating the achievements of exceptional individuals in this way obscures the fact that science is, in fact, a collective enterprise.

The anthropologist Hélène Mialet, who carried out an ethnographic study of Hawking's working practices in 2012, showed how unrealistic a description the 'lone genius' was in his case.[31] Mialet reveals Hawking to have been an active manager of a range of different material, technological, and human communities. It took a great deal of organisation to make sure that all the practical tools were at hand to facilitate his living and working. Bodily care depended on paid and family labour. In the pre-internet age, Hawking's access to the scientific community came in the form of bodily labour performed by students. They would sort and show him preprints of papers, pinning them to the wall so that he could decide whether he was interested in their contents. They would hold books and papers in front of him, turning the pages so he could read. This kind of labour became unnecessary once scientific journals moved online, but as Kip Thorne and others have pointed out, other forms of physical engagement with the world are still essential for theoretical physics. Scholarship in the humanities could potentially be expressed through a purely verbal format: physicists, however, need a board (digital or physical) to calculate, diagram, and communicate with their peers. Hawking was unable to write or draw in this way, but used diagrams created earlier in the century by Penrose and Feynman, in tandem with his powers of visualisation, to develop his understanding of the universe.[32]

Working closely with his graduate students, who did the physical work of calculation and diagramming, Hawking was able to internalise and build on their productions; the students interpreted and translated his responses to create the scientific papers that went into circulation under Hawking's name. Such a distribution of labour is fairly common in scientific research, even amongst non-disabled researchers. As Mialet noted, the only unusual aspect of this relationship is that the collaborative networks of discursive interaction on which science depends, and which are usually taken-for-granted or hidden, were made visible in this case because of Hawking's physical differences.

Mialet's insights did not reflect the popular picture of Hawking. In the aftermath of the publication of *A Brief History of Time* (1988), the media consistently positioned him in the lone genius mode of Einstein, Newton, and other scientific heroes. Hawking's response was blunt:

> It's very embarrassing. It's rubbish, just media hype. They just want a hero and I fill the role model of a disabled genius. At least, I am disabled, but I am no genius.[33]

At the same time, Hawking's self-constructed identity clearly drew on the heroic tradition. In some instances, this was clearly done with an eye to the audience. He noted, for example, that emphasising his connections with the 'father of computing', Charles Babbage, made fundraising for Cambridge University in California's Silicon Valley easier.[34] Babbage, like Hawking, had been appointed as Lucasian Professor of Mathematics at Cambridge, a position also once held by Isaac Newton. This led some media outlets, though not Hawking himself, to identify Hawking as the 'heir to Newton'. Hawking did use other scientific heroes to contextualise himself. He frequently pointed out that he was born in 1942, exactly three hundred years after the death of Galileo, and that it was 'shortly after' he himself had travelled as a young man to the Vatican to receive a medal from Pope Pius XII 'that the Church rehabilitated Galileo'.[35] Consciously or not, by drawing attention to these aspects of his biography, Hawking was positioning himself within the lineage of scientific stars that began with Copernicus, included Newton and thus far had ended with Einstein.[36]

As noted above, fragments of Hawking's desk were made over into expensive watches for buyers who wished to link themselves physically to his achievements. An auction of his possessions, including his wheelchair and doctoral thesis, raised around £1.8 million for charity.[37] Even within his own lifetime, objects associated with Hawking seemed to take on greater significance. British-born astronaut Piers Sellers took Hawking's Copley medal (awarded by the Royal Society in 2006) to the International Space Station.[38] Working in collaboration with the sculptor Eve Shepherd, Hawking helped to create a statue of himself, which would be exhibited at Cambridge University. The anthropologist Mialet was present when the plaster version of the sculpture was shown to Hawking and his colleagues for the first time. One colleague suggested it be placed in the corridor outside Hawking's office and enabled to 'speak' with Hawking's voice so that confused students could seek enlightenment from it. Unlike Newton's tutelary spirit, which could only watch silently over the assembled savants at his tomb, Hawking and his distinctive voice synthesiser could potentially become a speaking oracle for future generations.

Hawking's body, like that of Newton, Darwin, and many other scientists (including David Livingstone, the missionary-scientist we last saw in Chapter 3), was interred in London's Westminster Abbey, in June 2018. In May 2021, the Science Museum (London) took possession of the contents of his office.[39] This included his books and papers, the photographs and illustrations that had covered his walls, and the chalks for his blackboard; it also included his washing up bowls, his pill cutters, Allen keys, and orange ethernet cables. In her important essay on scientific heroes, Ludmilla Jordanova noted that 'the vast majority of objects in the collections of science museums relate far more directly to ordinary "business as usual" science than they do to the actions of the exceptional few'.[40] In Hawking's case, the Museum's collection certainly includes the material objects that encapsulate the networks of people and machines that made his work possible. Whether the significance of those communities to Hawking's achievements can be made plain to future audiences remains a question. As we have seen in relation to both historical and national genealogies of science, the aura which sets exceptional scientists apart does not emerge in and of itself but is actively created by audiences and institutions. Jordanova notes that museums are particularly powerful in lending an aura of sanctity to otherwise ordinary scientific ephemera.

Our wider history of scientific sanctification has highlighted the connection between the language and practice of hagiography and the emergence of the modern conception of genius. The only register initially available in which to understand the outstanding achievements of particular scholars was that of sainthood: by drawing on this repertoire, commentators were able to delineate and celebrate new scientific practices. As the modern category of genius gained cultural traction, it brought with it some of the theological tensions implicit in hagiography, including the question of whether spectacular creative innovation was the result of the divine spark of innate brilliance or the product of sustained labour and discipline. Specific individuals like Newton and Einstein were credited with the miraculous transformation of their respective fields; the example of Hawking, meanwhile, shows the extent to which an ability to reimagine the world depends on the existence of collaborative networks of support that are hidden behind the cultural icon of the 'lone genius'.

Do metaphors matter? As this chapter has shown, the language and practice of hagiographic exceptionalism *is* important because it is a significant aspect of the boundary work that separates scientists from ordinary people and science from society. This chapter has suggested that scientific relics play an ambivalent role, on the one hand pointing towards transcendent knowledge, while on the other hand emphasising the embodied humanity of the scientist. They are defined by the exceptionalism of their originator, while simultaneously enabling their audience to identify with him (or occasionally her). In this sense, relics encapsulate a tension at the heart of science communication: how to represent science as an exceptional human achievement while acknowledging that the political and economic impact of scientific developments is not evenly distributed across populations and environments.

In the following chapters, we will examine some strategies that can be used to broaden the options available within science communication and in relation to religion. We will look at new approaches that take storytelling seriously, which focus on the active participation and engagement of audiences and go beyond the Western focus on 'science-and-religion-at-war'. *The Big Bang Theory*'s Sheldon Cooper may have been militantly atheist, but his colleague, the astrophysicist Raj Koothrappali, regularly attended temple in order to affirm his sense that the universe operates according to a pattern that could be apprehended, even if not fully understood.

Further reading

Amy C. Chambers, Lisa Garforth, Miranda Jeanne, Marie Iossifidis, and Joanna Verran, *Reading Science/Fiction: Practices, Pleasures and Publics* (Singapore: Springer Nature Singapore, 2025)

Karl Giberson and Mariano Artigas, *Oracles of Science: Celebrity Scientists versus God and Religion* (New York: Oxford University Press, 2007)

Roslynn Haynes, *From Faust to Strangelove: Representations of the Scientist in Western Literature* (Baltimore: Johns Hopkins University Press, 1994)

David A. Kirby, *Lab Coats in Hollywood: Science, Scientists, and Cinema* (Cambridge, MA: MIT Press, 2011)

Notes

1. Neil deGrasse Tyson, https://twitter.com/neiltyson/status/548140622826459136?lang =en, 25 December 2015, accessed 23 April 2024.

2. For example, Richard Dawkins, 'Happy Newton Day!', *The New Statesman*, 13 December 2007, https://www.newstatesman.com/long-reads/2007/12/birthday-jesus-lady-god accessed 4 March 2025.

3. Campbell Books and Nila Aye, *Scientists: My First Heroes* (London: Campbell Books, 2020); Saskia Gwinn and Ana Albero, *Scientists Are Saving the World! So Who Is Working on Time Travel?* (London: Magic Cat Publishing, 2024); Maria Isabel Sanchez Vegara and Frau Isa, *Marie Curie: Little People, BIG DREAMS* (London: Frances Lincoln Children's Books, 2017); David Pakman, *Think Like a Scientist: A Kid's Guide to Scientific Thinking* (Independently published, 2023).

4. Kevin D Finson (2010) 'Drawing a Scientist: What We Do and Do Not Know After Fifty Years of Drawings', *School Science and Mathematics* 102.7 (2010): 335–345, https:// doi.org/10.1111/j.1949-8594.2002.tb18217.x.

5. The need to recruit more female scientists was the theme of one episode of *The Big Bang Theory* (season 6, episode 18, 'The Contractual Obligation Implementation'). See also Heather Mackintosh, 'Representations of Female Scientists in *The Big Bang Theory*', *Journal of Popular Film and Television*, 42.4 (2014): 195–204, https://doi.org/10.1080/ 01956051.2014.896779; Margarat A. Weitekamp, 'The Image of Scientists in *The Big Bang Theory*', *Physics Today*, 70.1 (2017): 40–48, https://doi.org/10.1063/PT.3.3427. See also Amy C. Chambers, 'Representing Women in STEM in Science-Based Film and Television', in *The Palgrave Handbook of Women and Science Since 1660*, edited by Claire G. Jones, Alison E. Martin, and Alexis Wolf (London: Palgrave, 2022): 483–501.

6. See, for example, the Royal Society's collection 'Equality, Diversity and Inclusion', n.d., https://royalsociety.org/current-topics/diversity/, accessed 23 April 2024; and Nikki Forrester, 'Diversity in Science: Next Steps for Research Group Leaders', *Nature* 585 (2020), S65–67, https://www.nature.com/articles/d41586-020-02681-y.

7. Anna Maerker, 'Hagiography and Biography: Narratives of "Great Men of Science"', in *History, Memory and Public Life: The Past in the Present*, edited by Adam Sutcliffe, Simon Sleight et al. (London: Routledge, 2018).

8. Thomas Carlyle, *On Heroes, Hero-Worship and the Heroic in History* (London: James Frazer, 1841); see also Ludmilla Jordanova, 'On Heroism', *Science Museum Group Journal* 1.01 (2014), unpaginated, https://dx.doi.org/10.15180/140107.

9. Owen Gingrich, 'The Copernican Quinquecentennial and its Predecessors', *Osiris*, second series, volume 14 (1999): 37–60, https://www.journals.uchicago.edu/doi/abs/10.1086/

649299; see also Thony Christie, 'Nicky was an Ermländer!', *The Renaissance Mathematicus*, (2009), https://thonyc.wordpress.com/2009/07/18/nicky-was-an-ermlander/, accessed 19 February 2025.

10. John Maynard Keynes, 'Newton, the Man', in *The Collected Writings of John Maynard Keynes* (Cambridge: Cambridge University Press, 1978); Daniel Kuehn, 'Keynes, Newton and the Royal Society', *Notes and Records: The Royal Society Journal of the History of Science*, 67.1 (2012): 25–36, https://doi.org/10.1098/rsnr.2012.0053.

11. Richard Yeo, 'Genius, Method and Morality: Images of Newton in Britain, 1760–1860', *Science in Context* 2.2 (1988): 257–284, doi:10.1017/S0269889700000594; Patricia Fara, *Newton: The Making of Genius*, (London: Macmillan, 2002).

12. Robert Iliffe, 'St Isaac: Newtonian Hagiography and the Creation of Genius', in *Savant Relics: Brains and Remains of Scientists*, edited by Maria Conforti Beretta, Paolo Mazzarello et. al. (Sagamore Beach, MA: Science History Publications, 2016).

13. Robert Iliffe, 'St Isaac: Newtonian Hagiography and the Creation of Genius', in *Savant Relics: Brains and Remains of Scientists*, edited by Maria Conforti Beretta, Paolo Mazzarello et. al. (Sagamore Beach, MA: Science History Publications, 2016), pp. 104–105.

14. Rebekah Higgitt, *Newton: Newtonian Biography and the Making of the 19th Century History of Science*, (Pittsburgh: University of Pittsburgh Press, 2016), p. 10.

15. Mark Lesney, 'Newton's Hair', *Chemistry Chronicles* (April 2003): 31–32, https://pubsapp.acs.org/subscribe/archive/tcaw/12/i04/pdf/403chronicles.pdf?, accessed 18 February 2025.

16. Richard Yeo, 'Genius, Method and Morality: Images of Newton in Britain, 1760–1860', *Science in Context* 2.2 (1988): 257–284, doi:10.1017/S0269889700000594, p. 260.

17. Thomas Chalmers, *Evidence and Authority of the Christian Revelation*, Edinburgh (1815).

18. Richard Yeo, 'Genius, Method and Morality: Images of Newton in Britain, 1760–1860', *Science in Context* 2.2 (1988): 257–284, doi:10.1017/S0269889700000594, p. 264.

19. Gerald Holton, 'On trying to understand scientific genius', *The American Scholar* 41.1 (1971): 95–110, https://www.jstor.org/stable/41209034, p. 97.

20. Gerald Holton, 'On trying to understand scientific genius', *The American Scholar* 41.1 (1971): 95–110, https://www.jstor.org/stable/41209034, p. 108, paraphrasing Einstein (1918).

21. Open University, 'Humphry Davy, Laughing Gas and the Era of Self-Experimentation', 2017, https://www.open.edu/openlearn/history-the-arts/history-science-technology-medicine/humphry-davy-laughing-gas-and-the-era-self-experimentationt, accessed 3 February 2025.

22. William Whewell, *Astronomy and General Physics Considered with Reference to Natural Theology* (London: William Pickering, 1833), p. 307.

23. William Whewell, *Astronomy and General Physics Considered with Reference to Natural Theology* (London: William Pickering, 1833), p. 308.

24. Marco Beretta, Maria Conforti, and Paolo Mazzarello, eds., *Savant Relics: Brains and Remains of Scientists* (Sagamore Beach, MA: Science History Publications, 2016).

25. John Holmes, *Temple of Science: The Pre-Raphaelites and Oxford University Museum of Natural History* (Oxford: Bodleian Library, 2020).

26. Robert D. Hicks, 'Marco Beretta, Maria Conforti, and Paolo Mazzarello (Editors), *Savant Relics: Brains and Remains of Scientists*', *Isis* 109.1 (2018): 154–155, p. 155.

27. Ludmilla Jordanova, 'On Heroism', *Science Museum Group Journal* 1.01 (2014), unpaginated, https://dx.doi.org/10.15180/140107.

28. Bremont Watches, 'Limited Edition Hawking: Unlocking Mysteries of the Universe', https://www.bremont.com/products/bremont-hawking-rose-gold, accessed 23 April 2024.

29. Museo Galileo, 'Savant Relics' (2015), https://www.museogalileo.it/en/library-and-research-institute/publications-and-conferences/conferences-and-workshops/661-savant-relics-brains-and-remains-of-scientists-4th-watson-seminar-in-the-history-of-material-and-visual-science-en.html, accessed 23 April 2024.

30. Alexandra Walsham, ed., *Relics and Remains, Past and Present* (Oxford: Oxford University Press, 2010).
31. Hélène Mialet, *Hawking Incorporated: Stephen Hawking and the Anthropology of the Knowing Subject* (Chicago: University of Chicago Press, 2012).
32. David Kaiser, 'Physics and Fenyman's Diagrams', *American Scientist*, 93 (2005): 156–165, https://web.mit.edu/dikaiser/www/FdsAmSci.pdf, accessed 18 February 2025.
33. Lisa Kremer, 'The Smartest Person in the World Refuses to be Trapped by Fate', *Brainstorm*, 1 August 2012, https://brainstrom.org/the-smartest-person-in-the-world-refuses-to-be-trapped-by-fate/, accessed 24 April 2024.
34. Hélène Mialet, *Hawking Incorporated: Stephen Hawking and the Anthropology of the Knowing Subject* (Chicago: University of Chicago Press, 2012), p. 137.
35. Hélène Mialet, *Hawking Incorporated: Stephen Hawking and the Anthropology of the Knowing Subject* (Chicago: University of Chicago Press, 2012), p. 99.
36. In June 1993, Hawking appeared in the TV series *Star Trek: The Next Generation*, playing poker with Newton, Einstein, and the show's android, Data ('Descent, Part One', season 6, episode 26). Although the show's storylines included many historical figures, Hawking was the only 'real-life' person to appear as himself.
37. BBC online, 'Stephen Hawking personal effects fetch £1.8 million at auction', 8 November 2018, https://www.bbc.co.uk/news/uk-england-cambridgeshire-461443632018, accessed 23 April 2024.
38. Tara Cox, 'There's Already a Steven Hawking statue', *Cambridge News*, 15 March 2018, https://www.cambridge-news.co.uk/news/cambridge-news/stephen-hawking-cambridge-memorial-statue-14411013, accessed 24 April 2024.
39. Science Museum Group, 'Stephen Hawking's Office', https://www.sciencemuseum.org.uk/stephen-hawkings-office, accessed 23 April 2024.
40. Ludmilla Jordanova, 'On Heroism', *Science Museum Group Journal* 1.01 (2014), unpaginated, https://dx.doi.org/10.15180/140107.

10

New atheism, new faith

Things can only get better

Over the twentieth century, science got better, people got smarter, and religious belief declined. Could it be mere coincidence? James Allan Cheyne, writing in the *Skeptic* in 2009, echoed the belief of many that these things were surely connected: 'people are getting smarter, smart people tend to reject irrational beliefs, hence with increasing intelligence more people become nonbelievers'.[1] Writing forty years earlier in the *New York Times*, the social theorist Peter Berger predicted that these trends would create a hostile environment for believers by the twenty-first century; they were 'likely to be found only in small sects, huddled together to resist a worldwide secular culture'.[2] The sociologist Steve Bruce has similarly argued that the growing influence of science will drive a process of secularisation leading to the decline of the influence of religion both on individual lives and society as a whole.[3]

At first glance, the data seem to bear out Bruce's predictions. The 2021 UK census revealed that for the first time in that survey, less than half the population (forty-six per cent) described themselves as 'Christian'.[4] This represented a thirteen per cent decrease from the 2011 census. The second-largest group within the population described themselves as of 'No religion' and had grown from twelve per cent in 2011 to thirty-seven per cent in 2021. This data came as no surprise to UK Christian churches, since the numerical decline of those attending worship in churches has been well-documented.[5] UK church membership has declined from 10.6 million in 1930 (thirty per cent of the population) to 4.8 million (8.5 per cent) in 2022. In 2023, less than five per cent of the English population attended church.[6]

It is strange that, just as the process of secularisation seemed to be secured, people started pleading all the more loudly and urgently for atheism. In the early twenty-first century, four men began to be spoken of as the 'four horsemen' of the 'new atheism'.[7] The title, an amusing nod to the agents of apocalypse in the book of Revelation, was yet another indication of the

Science, Religion, and the Human Future. Amanda Rees et al., Oxford University Press. © Amanda Rees, Franziska E. Kohlt, Tom McLeish, Charlotte Sleigh, David Wilkinson (2025).
DOI: 10.1093/9780191995316.003.0011

theological roots shared by Western science and Christianity. The neuro-scientist Sam Harris was first out of the blocks with his 2004 broadside *The End of Faith: Religion, Terror, and the Future of Reason.* Richard Dawkins, a biologist turned science writer, followed up with a 2006 TV series about religion entitled *The Root of All Evil?* (Plot spoiler: the question is answered in the affirmative.) The series was rewritten as *The God Delusion* (2006), which peaked at number four in the *New York Times* bestseller list and has been translated into thirty-one languages, selling over two million copies worldwide.[8] Over the coming years, journalist Christopher Hitchens and the philosopher and cognitive scientist Daniel Dennett added to the new atheist oeuvre, which took the form of books, broadcasts, and social media posts.

In this chapter, we contextualise the new atheism debates and examine what was at stake in them. We look at some of the Christian responses to them, noting the symmetry between the some of the theologies invoked by both sets of practitioners. We note some of the reasons for the fading of new atheism.

In the shadow of the towers

While it would be fair to say that the popularity of new atheism probably peaked in the 2010s, the ripples of its argument continue to influence Western culture in the 2020s.[9] Comedians such as Stephen Fry, Ricky Gervais, and Tim Minchin often reference new atheist thinking, keeping it alive in the public sphere. Other scientists have been encouraged to publicly identify as atheists, for example, Stephen Hawking, who became more explicit about his atheist belief in his book *The Grand Design* (2010).[10] New atheism is difficult to understand as a coherent philosophical position. It is better to think of it as a broad position made up of some common themes, not all of which would be held by all participants or expressed with equal force.

In the background, and indeed the foreground, of the new atheists' pronouncements were the horrific events of Tuesday, 11 September 2001. On that day, the Islamic terrorist group al-Qaeda carried out a series of four coordinated terrorist attacks on the United States. These attacks killed nearly 3000 people and destroyed the Twin Towers of New York's World Trade Centre. Their impact on US politics was profound, and they acted as a catalyst for the reshaping of global politics through a controversial and tragic war between the West and Iraq, as well as the invasion of Afghanistan.

Although the four horsemen had been raised in a context of cultural Christianity, Islam was often their target. Christopher Hitchens' book title *God Is Not Great* (2008) deliberately maligned the Muslim cry of faith uttered in all circumstances: *Allahu Akbar* (God is most great). Some new atheist sentiment took the trouble to distinguish between fundamentalist and mainstream Islam and some did not. Meanwhile, some white Christians felt that they were treated worse than Muslims by the new atheists, believing that the cultural sensitivities extended to what was, in Western context, a largely ethnic minority religion, were not accorded to them. Hitchens was pretty even-handed in his distributing his venom; his anti-Muslim title included St Francis of Assisi, Gandhi, Mother Teresa, and the Dalai Lama as examples of people whose goodwill was, at best, hampered by their faith, and at worst was a duplicitous façade.[11] Sam Harris' attacks were aimed not only at extremist or fundamentalist positions but also at moderate faiths such as Roman Catholicism and mainstream Protestantism. From his perspective, even moderate approaches contributed to the problems of the world by tolerating and teaching things that are false.[12]

The cultural authority of the new atheists was limited by their own homogeneity, although some iterations of the four horsemen did also include a fifth: the Somali-born writer and politician Ayaan Hirsi Ali. As a woman of colour, she apparently demonstrated that the men's distaste for Islam was not a matter of patriarchal and racist bias; her conversion to Christianity in 2023 caused consternation. Whether or not the new atheist movement could be characterised as Islamophobic, it sat comfortably within an influential philosophy that had been sketched by Samuel Huntington in 1992. Looking at the ruins of the Communist world at the end of the Cold War, Huntington judged that future conflicts would not be fought between nation states. Instead, the world could look forward to the 'clash of civilisations', with religious and cultural identity the cause of irreconcilable conflict. Covers of Huntington's book, which was eventually published in 1996, featured religious symbols. These emphasised the role of faith in conflict generation before the reader had even opened its pages. Huntington's division of the world was idiosyncratic, with some civilisations named by religion (Islamic, Hindu, Orthodox, Buddhist) and some by location (Latin American, African, Sinic, Japanese).[13] Huntington's own civilisation, and that of the four horsemen, eluded religious and geographical definition. It was called 'Western' and it included North America, half of Europe, and Australia and New Zealand.

There was a lot of criticism of the 'clash of civilisations' thesis. People argued that it was too monolithic and too deterministic.[14] Not every African thinks the same way, to put it mildly. In the context of the overlap with new atheism, these criticisms took on an additional dimension. So far, as many new atheists were concerned, Huntington's Western civilisation might as well have been called 'scientific'. The astrophysicist Adam Frank, for example, has proclaimed that 'to defend Western civilisation', we should 'start with science'.[15] If science is a historic product of Western culture, as Frank and many others believe, then it is alien to other cultures' way of thinking. 'They' just won't get it when 'we' tell them what is good for them. It sounds a little like the arguments we encountered in the chapter about AI, whereby some types of people are not intelligent enough to engage with the advances of technology.

The scientific credentials of the horsemen are very important to their position within the new atheism movement. Their arguments, too, have hinged on matters of science. Dennett, for example, argued that religion needs to be studied from a biological perspective in order to understand the experience of faith as a natural process which is based in the structure of the human brain and the evolution of the human body.[16] This may sound similar to the zoologist Alister Hardy's spiritually inclined arguments about the 'biology of God'. Hardy's concern, however, was to resist scientific reductivism and expand the horizons of scientists—not to use biology to explain religion away.

More fundamentally, the new atheists have portrayed science as proving religion wrong; it rules out any good faith possibility (if the pun will be excused) of religious belief. Their argument hinges on the idea that science is the best or only way of knowing about the world. This is often called 'the scientific method', a term that was introduced in the twentieth century. As we saw in relation to 'Climategate' in Chapter 6, the scientific method is described in various ways but often involves creating a hypothesis and testing it in repeatable experiments, usually in a laboratory. Because religion cannot use 'the scientific method', it is worthless. God is not proven.

Philosophers (mostly secular) quickly poked holes in this description of fact-getting.[17] We can observe something many times but we can never be sure that the next time will be the same.[18] Our frameworks of thinking and being in the world map the experiments that we construct and our interpretations of them: as Bruno Latour argued, rather than an experiment determining whether a hypothesis is right or wrong, you often need to know the right answer before you can determine the kind of experiment you need. Nevertheless, the myth about fact-getting in science has proved persistent.

The horsemen were hot on finding inconsistencies between scripture and science, and within scripture itself, that proved doctrine and theology to be factually incorrect. One example was the point that light appears in the Genesis 1 account of creation before the sun and moon are brought into being. Dawkins had his real eureka moment when he noticed something about the overall narrative of the Bible. In a now famous passage, he wrote:

> The God of the Old Testament is arguably the most unpleasant character in all fiction: jealous and proud of it; a petty, unjust, unforgiving control-freak; a vindictive, bloodthirsty ethnic cleanser; a misogynistic, homophobic, racist, infanticidal, genocidal, filicidal, pestilential, megalomaniacal, sadomasochistic, capriciously malevolent bully.[19]

This, he claimed (continuing the tradition of Thomas Paine's critiques of Biblical morality), is inconsistent with the New Testament's emphasis on a God of love. Theologians have been discussing this apparent contradiction for centuries and have in fact taken it much further than Dawkins; they find contradictory accounts of God even *within* the Old Testament. Theologians tend to see ambiguity and tension as an opportunity for fruitful reflection, but in the new atheists' world, something is either right or wrong. Dawkins' theology was thin; his God was a kind of evil Santa and any amendment of this description within scripture was heresy, notwithstanding God's nonexistence.

Dawkins took the apparent intellectual inconsistency of Christianity a step further. He argued that theology is not a real academic subject at all and should therefore be rendered in scare quotes and banned from university life:

> What has 'theology' ever said that is of the smallest use to anybody? When has 'theology' ever said anything that is demonstrably true and is not obvious?[20]

Dawkins used this argument (or perhaps 'argument'?) to object to the appointment of Professor Keith Ward, a philosophical theologian, to the Regius Professorship of Divinity at his own University of Oxford. In a robust exchange with Dawkins, Ward argued that new atheism seemed to have a deeply emotional antipathy to the idea of a moral and spiritual purpose for human life, rooted in a view of religion as anthropomorphic, literalistic, and antithetical to life, joy, and freedom. No doubt some people may experience faith, in this manner. But to characterise all religion in this way is to fail to make important discriminations between various kinds of belief in God.[21]

In some ways, the idea of science's superiority to religion reflects some of the earlier positions on social evolution and science which we considered in Chapter 2. As well as being foundational to the emergence of the secularisation hypothesis, Auguste Comte's positivist philosophy treated science both as the sole source of reliable knowledge and the highest stage in human social evolution. From this perspective, the only standards which count when it comes to assessing truth or falsity are those derived from empirical judgement. But there are also important differences from these earlier assertions of scientific superiority, as the theologian John Haught has argued.[22] While the horsemen celebrated atheism and anti-theism as the basis for human liberation, earlier philosophers like Camus, Sartre, and Nietzsche were deeply concerned about the cost of denying God's existence: what would give human life meaning in the absence of religion? In contrast, Dawkins affirmed: 'You can be an atheist who is happy, balanced, moral, and intellectually fulfilled'.[23]

It is a mistake to imagine that religion is in the business of trying to prove God exists. Faith, from the perspective of the believer (or of the sociologist, for that matter), means something very different to what the horsemen assumed. As we saw in the chapter on climate change, it is more about imaginative doing than believing. Dawkins sticks with defining religion in terms of belief, which he then sets out to disprove from within his scientistic framework. He gives a standard critique of the three traditional arguments for the existence of God: the cosmological, the ontological, and the argument from design. In so doing, he ignores one of the central tenets of the Abrahamic religions (Judaism, Christianity, and Islam): you cannot prove the existence or nature of God because God is self-revelatory. For Christians, God's love is revealed by Jesus of Nazareth, whom they believe was God. This is not the same thing as Jesus proving the existence of God. While some Christians have sought to prove the existence of God, that proof has historically been irrelevant to their faith. We saw, for example, in Chapter 1, how Anselm of Canterbury used logic to demonstrate God's existence in his *Proslogion*; and that in doing so, he was, quite literally, preaching to the converted.

Unsurprisingly, many new atheist claims have been anchored by reference to the conflict narrative, treating Galileo and Darwin as heroes who were persecuted by religious institutions for demonstrating that the Bible is factually incorrect. As earlier chapters of this book have shown, this reading of historical events is misleading and lacks support from historians of science, but this did not dim the enthusiasm of new atheism for these stories. Dawkins even identified himself with the beleaguered martyrs of olden times, writing that his aim in *The God Delusion* was to give fellow atheists the confidence to 'come out of the closet'. From Dawkins' perspective, religion enjoys undue power and privilege: his aim is to give those outside its tent the intellectual resources and leadership to fight discrimination and make space for atheism in the contemporary world. His turn of phrase, 'coming out of the closet', is an interesting and troubling choice. It is most often used to describe the danger of disclosing one's homosexuality in a homophobic culture. Dawkins appears to suggest that he is in a persecuted minority. For those who hold to the 'civilisational threat' thesis, such language can be a code for racist or Islamophobic anxiety. It is, in truth, difficult to imagine a well-educated white man such as Dawkins being barred access to the law, to government, to university leadership, on account of an atheist creed.

Symmetrical fundamentalisms

New atheists and some evangelical Christians expended their energies in mutual denigration, using remarkably similar methods.[24] Demonisation might not even be too strong a word for their mutual portrayals. A conflict model is not just a way of simplifying the world or establishing an identity as distinct from the other; it can be a dangerous source of injustice. Just in case Christian theologians point the finger too much at Huxley, Dawkins, and others for their use of the conflict model, it is wise to acknowledge the troubling role of conflict theses (plural) within Christian theology and mission.

The position of scientific creationism, for example, employs the same conflict model used by atheistic scientism, starting from the premise that science is in conflict with religion. In the case of six-day and young Earth creationists, it is a particular interpretation of Genesis which is sent into warfare with the majority of mainstream science. Young Earth creationists are often keen to demonstrate that they can do science just as well, if not better than, the scientists. They use this to build up their theories,[25] but even more to knock down the findings of conventional geology. They possess an extraordinary knowledge of scientific detail which they deploy to find holes and cracks in agreed science, with the apparent assumption that any shades of grey in its evidentiary basis must rule it out altogether. There are clear parallels here with the way that evidence for anthropogenic climate change has been treated by climate sceptics and fossil fuel lobbyists.

Even mainstream Christian apologetics of the twentieth and twenty-first centuries has often been practised in a spirit of scientific method: empiricism and rationality. A common trope suggests that inquirers to the faith deploy rational methods to decide whether Jesus was 'mad, bad, or who he said he was'. The most famous twentieth-century example of scientific apologetics is arguably provided by *Who Moved the Stone?* This book, first published in 1930 and reprinted numerous times, marshalled empirical evidence and rational argument to adjudicate on the hypothesis of Jesus' resurrection. Famously, its author claimed to have begun in a spirit of debunking, only to be persuaded by the evidence. G. K. Chesterton compared it to a detective narrative, at a time when crime investigation too was transforming into a science. There was an answer, if one pursued it honestly and intelligently through the evidence. In the realms of more academic theology, Alister McGrath suggests that a Christian apologist can 'usefully draw upon science's own methods', allowing him to infer the best explanation from the evidence. Christ's resurrection hovers in the background as the paradigmatic case for this exercise. Although McGrath would not dream of reducing Christian theology to a scientific method, he does sense a natural affinity between the two. He highlights the Christian's 'health and balance ... sense and vigour' in thinking, which seems not too far from the empirical ideal.[26] It is sometimes observed by critics of science fiction films that the 'good guys' often come to resemble their enemies, and the same might be the case when it comes to the conflict thesis; some Christians, at least, evince the epistemic virtues of science in defence of faith.

Another consequence of the deployment of the conflict narrative is the characterisation of the 'other' as suspect or dangerous. Six-day creationism has become in some parts of the Christian church a test of orthodoxy. (In other parts of the church, the same can be said for beliefs about sexuality.) It is a test of whether you are a true, Bible-believing Christian. Symmetrically, a belief in evolution (the field in which Dawkins did his research) is a quick badge to establish identity in terms of the professionalism of science. Here, we see the conflict narrative used to establish identity and to demonise the other. That is, if you do not hold to the six-day account in opposition to 'atheistic' evolution, then you are suspect or dangerous as a Christian. On the other side, those who do not adhere to (an often-reductionist version of) the theory of evolution find themselves beyond the pale of modern intellectual life.

Identity is formed by distancing from the other, and its use can bring with it sexism and racism. In a more serious case than six-day creationism, Christians have historically aligned their identity on antisemitic lines. The root of this prejudice and exercise of discriminating power can be attributed to numerous factors including competition between church and synagogue and the practice and theology of Christian missionaries. However, one primary factor has been the difficulty of holding both continuity and discontinuity between two religious movements stemming from a common root. The apostle Paul struggled to hold these together as the church grew out of Jewish communities, gentile converts, and the writings of early church theologians. In establishing Christian identity, it was very easy for some Christians to develop a conflict narrative between Jesus and the Jews, highlighting clashes that Jesus had with the Jewish religious authorities and attributing his death to them—and, by implication, to all Jewish people. From these beliefs followed the horror and violence of antisemitism on the part of 'Christian' Europe.

This form of the conflict narrative was challenged in a major movement of biblical scholarship known as 'The New Perspective on Paul'. In his book *Paul and Palestinian Judaism* (1977), E. P. Sanders argued that the picture of Judaism drawn from Paul's writings was historically false.[27] In their reading of Paul's letters, Christians had come to accept a view of Jews as mechanical followers of a futile law.[28] Projected back upon the gospels, this reading understood the Pharisees as metonymic for Judaism as a whole: hypocrites and persecutors.[29] Other scholars fingered Martin Luther as largely responsible for the negative readings of Judaism in Pauline writings. Working through issues of guilt and his struggles with Catholicism, Luther projected them onto the Jews and fuelled antisemitism for centuries to come.[30] It would be interesting to consider what issues the four horsemen—alike in many ways—were working through as they pinned the badge of enmity on Christianity or Islam.

It has been pointed out that Sanders did not go as far as he could have done in his revisionism, and 'remained more impressed by the *difference* between Paul's pattern of religious thought and that of first-century Judaism'.[31] This illustrates how the need to reject one conflict narrative can quickly lead to a new one being generated. Indeed, in the context of the science/religion conflict, many Christian writers get drawn into conflict with the conflict narrative!

Interdisciplinary work in science and theology

Alongside the development of 'scientific' apologetics, other theologians engaged constructively with science in their work. This work extends back in time to the 1960s, long before the coming of the four horsemen. We might perhaps see the work as a response to liberal theology of the 1960s, loosely clustered around an antagonism towards supernaturalism and thus apparently friendly towards science.[32] At the same time, theologians were coming to grips with the specific manifestation of science in the atomic and latterly the hydrogen bomb.[33]

A notable and important further development of the latter twentieth century was the emergence of 'scientist-theologians'.[34] (We have already seen, in Chapters 2 and 3, the longer tradition of this activity.) These researchers, who moved from scientific careers to writing as Christian theologians, have helped to create a new interdisciplinary area which has found expression both within popular culture and theological training. Key examples include the physicist Ian Barbour, the biologist Arthur Peacocke, and the mathematical physicist John Polkinghorne. There are other scientists from this era who, though not so intensive in their theological work, help to indicate a more substantial trend. The cosmologist Paul Davies and the astronomer Fred Hoyle, for example, saw their research as an opportunity to engage with theology and philosophy.[35] The astronomers Bernard Lovell, Rod Davies, Jocelyn Bell Burnett, and Arnold Wolfendale, all celebrated for their science, treated science as integral to their Christian faith. The same can be noted of the neuroscientist Donald MacKay, the geneticist Sam Berry, the physicist Andrew Briggs, the geologist Bob White, and the climate scientist John Houghton.

Polkinghorne's trilogy of books on science and theology, published in the late 1980s, was extremely influential in the United Kingdom in particular.[36] He identified continuities between the methodologies of both approaches, treating them equally as critical realist activities aimed at provisionally describing a common reality. Rather than trying to set out an overarching integration of science with theology, he preferred to begin with specific case studies from science, such as quantum theory, the Big Bang, or the end of the universe. From here, he would explore the theological consequences for understanding God and God's action in the world. Sometimes, these consequences offered confirmation of established theological positions and sometimes they represented challenges to those positions that required careful consideration.[37] His approach drew parallels with earlier efforts to develop feminist and liberationist theologies which similarly often began with specific examples. It was popular with lay Christians as well as those who did not identify directly with faith communities; as Polkinghorne said, more people were likely to be attracted to a talk on God and the end of the universe than a lecture on the philosophical basis of science.[38] Underpinning all of this work was the Anglican theological tradition of holding scripture, tradition, and reason together in the belief that the Creator God had spoken and acted supremely in the life, death, and resurrection of Jesus of Nazareth. It was from this basis that Polkinghorne went on to talk about how one should understand God at work in the world. He used complex chaotic systems such as the weather to illustrate his argument. Weather systems operate according to the known laws of physics but are not predictable in the way that the mechanistic Newtonian universe might suggest. For Polkinghorne, who made a strong link between epistemology and ontology, this meant that the Universe was open: the outcome of physical processes is not fully determined within Creation just as human history is not predestined, on account of human freedom.

Two names often cited alongside Polkinghorne are Alister McGrath and Tom McLeish. Alister McGrath, whom we have already met in this chapter, began his career as a biologist before being awarded a personal chair in theology at Oxford.[39] Tom McLeish worked throughout his career in physics, a vocation which culminated in a chair in Natural Philosophy at the University of York. This title, created especially for him, reflected his aspiration to recover the affordances accorded to such historical figures as Robert Grosseteste, a polymath able to flourish as an integrated investigator, philosopher and theologian.[40] McLeish continued active research into both science and science-engaged theology (SET), in part via the 'Ordered Universe' project which informs Chapter 1, until serious illness forced him into early retirement in spring 2022. His work brought science into conversation with the biblical wisdom tradition, developing insights into creation and the reconciling role of both science and theology, emphasizing the creative, imaginative elements of scientific research, and their resonance with work in theology, music, and poetry.[41] Not least in his role as Chair of the Royal Society's Education Committee, he took a leading role in developing new strategies for interdisciplinary research and education and worked extensively and publicly to challenge the conflict narrative.

Andrew Perry and Joanna Leidenhag have refreshed Polkinghorne's legacy, showing that authentic interdisciplinary engagement comes when questions are asked and answered in good faith. Theology, they stress, can be done well by listening to and engaging with empirical science.[42] Leidenhag has defined the field as SET, a term which is gaining traction across universities in the United Kingdom. For her, it is a distinctive methodology, while others use the term more loosely to designate a collection of interdisciplinary activities.[43]

Seeds require good soil to grow, and it would be an incomplete description of interdisciplinary science and theology that overlooked the economic role of the John Templeton Foundation in encouraging and promoting it. Established by the British-American banker Sir John Templeton, the foundation was established to support work in the area of spirituality; it quickly became known for its special interest in the intersection of religion and science. Templeton stated: 'science and rationality are enemies not of religion—only of dogmatism'.[44] His foundation remains an unusually generous supporter of humanities research and, in recent years, has been a major supporter of projects implicitly or explicitly branded as SET, increasing its impact in the academic world and beyond.

The rise of the 'scientist-theologians' and the generous funding of their work have assisted a worldwide growing interest in the engagement of science and theology. Polkinghorne's 'bottom up' approach has meant that ideas and debates have flowed from scientists to theologians, and vice versa.[45] The voices of professional scientists have been heard within the humanist academy and the church. *Christians in Science* and *The Society of Ordained Scientists* reflect this trend, as well as the schools' initiatives *God and the Big Bang* and *Epistemic Insight*. At the fifteenth worldwide Lambeth Conference of Anglican Bishops (2022), one of the seven central themes was science.

What happened to new atheism?

The projections of twentieth-century sociologists and the hopes of the new atheists have not come to pass. For one thing, the data on rising atheism is not clear cut, despite the fall in church attendance. Work by the Theos Think-tank revealed considerable diversity within the category of Britons who stated their religious belief in the 2021 census as 'none'.[46] Only half of those who identified as 'Nones' were prepared to say that they did not believe in God. In fact, and perhaps surprisingly, nine per cent expressed a belief in God, alongside a further fourteen per cent who asserted the existence of a higher power. Indeed, an earlier 2018 study commissioned by the Christian charity Tearfund suggested that over half of all adults in the United Kingdom pray, and that they are increasingly likely to call on God while engaged in activities such as cooking or exercising. Just under half of those who pray said they believed God hears their prayers. Among the non-religious, personal crisis or tragedy was the most common reason for praying, with one in four saying they prayed to gain comfort or feel less lonely.[47] A follow-up survey in the midst of the 2020 pandemic found that more than two in five UK adults pray, including a quarter of adults who said they pray regularly (i.e., at least once a month or more often).[48] The falling numbers of self-identifying Christians in surveys are also balanced by a rise in those who describe themselves as 'spiritual but not religious'.[49] It may be that 'No Religion' is a means by which people can distinguish themselves from formal faith communities or indicate that they do not have a firm set of beliefs. In either case, an increase in the proportion of the UK populations who identify with 'No Religion' cannot be immediately taken to imply either the growth of secularism or the rise of atheism.

For another thing, recent sociological research reveals that religious faith is alive and well amongst practising scientists. A 2009 survey of members of the American Association for the Advancement of Science found that thirty-three per cent of scientists believe in God and another eighteen per cent in a higher power.[50] Religious faith seemed to vary according to disciplinary specialisation and age. Chemists seemed to be the most religious (forty-one per cent); forty-two per cent of younger scientists (aged 18–34 years) believed in God compared to twenty-eight per cent of those over sixty-five years of age. In a series of major global research projects, the sociologist Elaine Howard Ecklund has added considerably to our knowledge of the relationship between scientists and faith in many different contexts.[51] In one study of 300 scientists at twenty-one elite US universities, she established that only a minority of scientists see religion and science as always in conflict. In another project, she found that the global picture was very different to the Anglo-American context. Examining the beliefs of scientists around the world, she found that while scientists in many regions were more secular than the general population, in others, around half of scientists see themselves as religious. She has also explored the heterogeneous nature of atheism in this context. More nuanced than monotonous stories of increasing secularisation, her work has documented the different pathways that researchers have taken, the diverse views of religion they hold, their perspectives on the limits to what science can explain, and their views of meaning and morality.

In 2022, researchers Nick Spencer and Hannah Waite carried out a qualitative survey of perceptions of science and religion in the UK context.[52] They conducted over a hundred in-depth expert interviews with scientists, philosophers, and sociologists, as well as commissioning a YouGov survey of more than 5000 UK adults. They concluded that both science and religion are highly complex, contestable, 'polyvalent' terms for their participants, meaning that the context in which their respondents encountered debates on these topics had a significant impact on the views that they were willing to express. Spencer and Waite concluded that within the UK setting, public perspectives on the science and religion debate have often been distorted by being viewed primarily through a few 'conflictual' lenses such as evolution vs creationism, the Big Bang vs God, and neuroscience vs religious experiences. The result is that communities find themselves in a situation where the idea of a 'war' between science and religion sits uncomfortably with their own lived experiences or hopes for the future. The Nigerian communities studied by Bankole Falade and Martin Bauer adopt a 'polyphasic' position of faith in both science and God.[53]

It seems then that the new atheism was flying against cultural headwinds. Secularism was more complicated than the sociological speculators had anticipated. It can mean the removal of divine discourse from public spaces, or even a decline in religious practice. But that is not the same as a society in which belief in God is a counter-cultural commitment.[54] Peter Berger and others have even proposed that what we are now seeing is the 'desecularisation' of the world, the resurgence and flourishing of religious belief in a scientific culture.[55] There were also specific reasons for the short-lived nature of the horsemen's project, to do with generational differences in culture. Sociologists have suggested that the project, at its outset, was marked by the broadly liberal politics of its instigators.[56] It chimed with younger people who were concerned about the discriminatory effects of religious dogma, on gender and sexuality, for example. These Millennials, however, came to critique the movement itself on the same grounds. They found its spokesmen too dogmatic, especially on multicultural themes.[57]

The famous dialogue between Archbishop Rowan Williams and Dawkins at Oxford University in 2012 certainly makes striking viewing when we consider the generational positioning of the new atheism debate.[58] No doubt intended as a rematch of the legendary Huxley–Wilberforce debate, it impresses the viewer for its genial, Senior Common Room vibe. Chaired by the emollient Sir Anthony Kenny, it is a most gentlemanly disagreement between white men of almost identical social background. It is hard to see the relevance of the conversation to most people in the United Kingdom.

In the next chapter, we will see how trends beyond the walls of Oxford's Sheldonian Theatre allow us to place the enhanced integration of science and faith in a broader context than the new atheists and their theological counterparts.

Further reading

Amarnath Amarasingam, ed., *Religion and the New Atheism: A Critical Appraisal* (Leiden: Brill, 2010)

Tom McLeish, *Faith and Wisdom in Science* (Oxford: Oxford University Press, 2016)

Samir Okasha, *Philosophy of Science: A Very Short Introduction*, 2nd edn (Oxford: Oxford University Press, 2016)

Notes

1. James Allan Cheyne, 'Atheism Rising: The Connection between Intelligence, Science, and the Decline of Belief', *Skeptic* 15.2 (2009): 33–38, gale.com/apps/doc/A2110 61622/AONE?u=anon~7fbe6cc9&sid=googleScholar&xid=491c95f4. Cheyne goes on to nuance these generalisations somewhat within his article. Alert readers of the earlier chapter on AI may query the notion of intelligence as a historically improving characteristic of humanity, and for that matter with the grand narrative of scientific progress.
2. Peter Berger, 'A Bleak Outlook Is Seen for Religion', *The New York Times*, 25 February 1968, p. 3.
3. Steve Bruce, *Religion and Modernization: Sociologists and Historians Debate the Secularization Thesis* (Clarendon: Oxford, 1992); Steve Bruce, *Religion in the Modern World: From Cathedrals to Cults* (Oxford: Oxford University Press, 1996); Steve Bruce, *God is Dead: Secularization in the West* (Oxford: Blackwell, 2002).
4. Office for National Statistics (ONS), statistical bulletin released on 29 November 2022, https://www.ons.gov.uk/peoplepopulationandcommunity/culturalidentity/religion/bulletins/religionenglandandwales/census2021, accessed 18 February 2025.
5. Peter Brierley, ed., UK Church Statistics No.4 (2021), https://www.brierleyconsultancy.com/shop/church-statictics-no-4-2021-digital-download, accessed 18 February 2025.
6. With the rise of streaming facilities for services, counting attendance can be difficult.
7. Christopher Hitchens, Richard Dawkins, Sam Harris, and Daniel Dennett, *The Four Horsemen: The Conversation That Sparked an Atheist Revolution* (London: Penguin Random House, 2019). The original discussions between the four men, which took place on 30 September 2007, were filmed and can be found on YouTube at https://www.youtube.com/watch?v=9DKhc1pcDFM, accessed 18 February 2025.
8. Richard Dawkins, *A Devil's Chaplain* (London: Weidenfeld & Nicholson, 2003); Richard Dawkins, *The God Delusion* (London: Bantam Press, 2006).
9. Elizabeth Bruenig, 'Is the New Atheism Dead?', *The New Republic*, 12 July 2015, https://newrepublic.com/article/123349/new-atheism-dead, accessed 18 February 2025.
10. Stephen Hawking and Leonard Mlodinow, *The Grand Design* (London: Bantam Books, 2010). A measured set of theological statements in *A Brief History of Time*, including concerns about arguments used to demonstrate a god of deism, hardened to a more dismissive position towards theistic belief in Hawking's later writing.
11. Christopher Hitchens, *God is Not Great: How Religion Poisons Everything* (London: Atlantic, 2008).

12. Sam Harris, *The End of Faith: Religion, Terror, and the Future of Reason* (New York: W. W. Norton, 2004); Sam Harris, *Letter to a Christian Nation* (London: Bantam, 2007); Sam Harris, *The Moral Landscape: Sow Science Can Determine Human Values* (London: Black Swan, 2012).

13. Samuel P. Huntington, *The Clash of Civilizations and the Remaking of World Order* (New York: Simon & Schuster, 1996), pp. 26–27.

14. Doubling down on Islamic fears, in *21 Lessons for the 21st Century* (London: Jonathan Cape, 2018), Yuval Noah Harari wrote that Huntington was too lax on Islamic fundamentalism, which was represented not just a clash with the West but a threat to the entire world.

15. Adam Frank, 'To Defend Western Civilization, Start with Science, Cosmos and Culture', NPR, 18 July 2017, https://www.npr.org/sections/13.7/2017/07/18/537882769/to-defend-western-civilization-start-with-science, accessed 18 February 2025. Frank gives brief acknowledgement to 'Muslim', Indian and Chinese scientific achievements, but states that the true 'scientific method' is a Western invention.

16. Daniel C. Dennett, *Breaking the Spell: Religion as a Natural Phenomenon* (New York: Viking, 2006).

17. Samir Okasha, *Philosophy of Science: A Very Short Introduction*, 2nd edn (Oxford: Oxford University Press, 2016).

18. One popular variant on the scientific method in the light of this point is the idea that science is in the business of trying to weed out bad hypotheses, while remaining modestly silent about what is correct. For a discussion of this issue, see Charlotte Sleigh, 'The Abuses of Popper', *Aeon*, 16 February 2021, https://aeon.co/essays/how-popperian-falsification-enabled-the-rise-of-neoliberalism, accessed 1 September 2024.

19. Richard Dawkins, *The God Delusion* (London: Bantam Press, 2006), p. 31.

20. Richard Dawkins, 'Scientific Versus Theological Knowledge', Letters, *The Independent*, 20 March 1993.

21. Keith Ward, *Why There Almost Certainly is a God* (London: SPCK, 2008).

22. John F. Haught, *God and the New Atheism: A Critical Response to Dawkins, Harris, and Hitchens* (Louisville: Westminster John Knox Press, 2008).

23. Richard Dawkins, *The God Delusion* (London: Bantam Press, 2006), p. 23.

24. William A. Stahl, 'One-Dimensional Rage: The Social Epistemology of the New Atheism and Fundamentalism', in *Religion and the New Atheism*, edited by Amarnath Amarasingam (Leiden: Brill, 2010), 95–108, https://doi.org/10.1163/ej.9789004185579.i-253. 38.

25. Timothy H. Heaton, 'Recent Developments in Young-Earth Creationist Geology', *Science & Education* 18 (2009): 1341–1358, https://doi.org/10.1007/s11191-008-9162-6.

26. Alister E. McGrath, 'The Natural Sciences and Apologetics', in *Imaginative Apologetics: Theology, Philosophy and the Catholic Tradition*, edited by Andrew Davison (London: SCM Press, 2011), p. 142.

27. E. P. Sanders, *Paul and Palestinian Judaism: A Comparison of Patterns of Religion* (London: SCM, 1977).

28. W. D. Davies, *Paul and Rabbinic Judaism: Some Rabbinic Elements in Pauline Theology* (London: SPCK, 1948).

29. Hyam Maccoby, 'Scribing the Pharisees', *London Review of Books* 13.9 (1991), https://www.lrb.co.uk/the-paper/v13/n09/hyam-maccoby/scribing-the-pharisees, accessed 18 February 2025.

30. Krister Stendahl, 'The Apostle Paul and the Introspective Conscience of the West', *Harvard Theological Review* 56 (1963): 78–96, doi:10.1017/S0017816000024779; G. F. Moore, *Judaism in the First Centuries of the Christian Era* (Harvard: Harvard University Press, 1927–1930, Vol. 2), p. 95.

31. James D. G. Dunn, 'The New Perspective on Paul', *Bulletin of the John Rylands Library* 65 (1983): 95–122, https://doi.org/10.7227/BJRL.65.2.6, p. 97.

32. The generation of British theologians writing on science included Charles Raven, Ian Ramsey, and Tom Torrance.

33. Dianne Kirby, 'Responses within the Anglican Church to Nuclear Weapons: 1945–1961', *Journal of Church & State* 37.3 (1995): 599–622, https://www.jstor.org/stable/23921096.
34. John C. Polkinghorne, *Scientists as Theologians: A Comparison of the Writings of Ian Barbour, Arthur Peacocke & John Polkinghorne* (London: SPCK, 1996).
35. Paul Davies, *The Mind of God: Science and the Search for Ultimate Meaning* (London: Penguin, 1993); Fred Hoyle, *The Intelligent Universe* (London: Michael Joseph, 1983).
36. John C. Polkinghorne, *One World: The Interaction of Science and Theology* (London: SPCK, 1986); John C. Polkinghorne, *Science and Creation: The Search for Understanding* (London: SPCK, 1988); John C. Polkinghorne, *Science and Providence: God's Interaction with the World* (London: SPCK, 1989).
37. John C. Polkinghorne, *The Faith of a Physicist: Reflections of a Bottom-Up Thinker* (Princeton: Princeton University Press, 2014).
38. John C. Polkinghorne, *Theology in the Context of Science* (London: SPCK, 2008).
39. Alister McGrath and Joanna Collicutt McGrath, *The Dawkins Delusion? Atheist Fundamentalism and the Denial of the Divine* (London: SPCK, 2007); Alister McGrath, *Christian Theology: An Introduction* (Oxford: Blackwell, 1998); Alister McGrath, *The Order of Things: Explorations in Scientific Theology* (Oxford: Blackwell, 2006).
40. Ordered Universe project, n.d., https://ordered-universe.com/, accessed 18 February 2025.
41. Tom McLeish, *Faith and Wisdom in Science* (Oxford: Oxford University Press, 2016); Tom McLeish, *The Poetry and Music of Science: Comparing Creativity in Science and Art* (Oxford, Oxford University Press, 2019).
42. John Perry and Joanna Leidenhag, 'What is Science-Engaged Theology?', *Modern Theology*, 37:2 (2021), 245–253, https://doi.org/10.1111/moth.12681, p.252.
43. Joanna Leidenhag, 'Science-Engaged Theology', St Andrews Encyclopaedia of Theology, 2024. Edited by Brendan N. Wolfe et al. https://www.saet.ac.uk/Christianity/ScienceEngagedTheology, accessed 11 February 2025.
44. John Templeton, *Possibilities for over One Hundredfold More Spiritual Information: The Humble Approach in Theology and Science.* Philadelphia: Templeton Foundation Press, 2000; John Templeton, and Robert L. Herrmann. *Is God the Only Reality?: Science Points to a Deeper Meaning of the Universe.* New York: Continuum, 1994.
45. See John Perry and Joanna Leidenhag, 'What is Science-Engaged Theology?', *Modern Theology*, 37:2 (2021), 245–253, https://doi.org/10.1111/moth.12681. See also Andrew Davison, 'Science-Engaged Theology Comes to San Antonio: A Report from the American Academy of Religion / Society of Biblical Literature Meeting 2021', *Theology and Science*, 20:1 (2022), 1–3, https://doi.org/10.1080/14746700.2021.2012913.
46. Hannah Waite, 'The Nones: Who They Are and What They Believe', Theos, 2022, https://www.theosthinktank.co.uk/cmsfiles/The-Nones---Who-are-they-and-what-do-they-believe.pdf, accessed 18 February 2025.
47. Savanta, 'Tearfund—Prayer Survey', n.d, https://savanta.com/knowledge-centre/poll/tearfund-prayer-survey/, accessed 18 February 2025.
48. Savanta, 'Tearfund COVID-19 Prayer Public Omnibus Research', n.d., https://savanta.com/knowledge-centre/poll/tearfund-covid-19-prayer-public-omnibus-research/, accessed 18 February 2025.
49. Nancy T. Ammerman, 'Spiritual but Not Religious? Beyond Binary Choices in the Study of Religion', *Journal for the Scientific Study of Religion* 52.2 (2013): 258–278, https://doi.org/10.1111/jssr.12024.
50. Pew Research Centre, 'Scientists and Belief', 2009, https://www.pewresearch.org/religion/2009/11/05/scientists-and-belief/, accessed 19 February 2025.
51. Elaine Howard Ecklund, *Science Vs. Religion: What Scientists Really Think* (New York: Oxford University Press, 2010); Elaine Howard Ecklund, *Secularity and Science: What Scientists around the World Really Think about Religion* (New York: Oxford University Press, 2019); Elaine Howard Ecklund, 'Science and Religion in (Global) Public Life: A Sociological Perspective', *Journal of the American Academy of Religion* 89.2 (2021): 672–700; Elaine Howard Ecklund and David R. Johnson, *Varieties of Atheism in Science* (New York: Oxford University Press, 2021).

52. Nick Spencer and Hannah Waite, "'Science and Religion': Moving Away From the Shallow End', (London: Theos, 2022), https://www.theosthinktank.co.uk/cmsfiles/Science-and-Religion-Moving-away-from-the-shallow-end_REPORT.pdf, accessed 18 February 2025.
53. Bankole A Falade and Martin W. Bauer, "'I Have Faith in Science and in God": Common Sense, Cognitive Polyphasia and Attitudes to Science in Nigeria', *Public Understanding of Science* 27.1 (2018), 29–46, https://doi.org/10.1177/0963662517690293.
54. Charles Taylor, *A Secular Age* (Cambridge, MA: Harvard University Press, 2007), p. 3.
55. Peter L. Berger, *The Desecularization of the World: Resurgent Religion and World Politics* (Washington, DC: Ethics and Public Policy Center, 1999).
56. Marcus Schulzke, 'The Politics of New Atheism', *Politics and Religion* 6.4 (2013): 778–799, doi:10.1017/S1755048313000217.
57. Amarnath Amarasingam and Melanie Elyse Brewster, 'The Rise and Fall of the New Atheism: Identity Politics and Tensions within US Nonbelievers', *Annual Review of the Sociology of Religion* (2016): 118–36, 10.1163/9789004319301_008, p.119. See also Stuart McAnulla, Steven Kettell, and Marcus Schulzke, *The Politics of New Atheism* (New York: Routledge, 2018).
58. Sir Anthony Kenny chaired a dialogue at Oxford University between Archbishop Rowan Williams and Professor Richard Dawkins on the subject of 'The Nature of Human Beings and the Question of Their Ultimate Origin'. The event was held on Thursday 23 February 2012 in the Sheldonian Theatre. 'Dialogue with Richard Dawkins, Rowan Williams and Anthony Kenny', https://youtu.be/bow4nnh1Wv0?si=PjCwK8dHhh_7WSZ9, accessed 18 February 2025.

11

New stories

Rethinking science

The new atheists, as we saw in the last chapter, presented science as the infinitely more robust alternative to faith as a source of knowledge. Theologians, as we also saw, renewed the historic trend of integrating these two fields. Unbeknown to these theologians, at least in the 1980s, they had allies elsewhere. These allies were scholars in Science and Technology Studies (STS), including historians, philosophers, and sociologists of science, all of whom were at work critiquing the scientistic assumptions of the 'four horsemen'. Scientism is the notion that science provides the best or only way of discovering truth. If there is not already a scientific answer to a question, there will be one day; and if there is no prospect of such an answer, then it is probably not a meaningful question. Across all its constituent disciplines, STS rejects the notion that science can provide answers of the sort that the scientistic atheists endorsed.

In the last chapter, we briefly reviewed the critique of empiricism, the idea that observations in experiments can tell us the 'laws of nature'. Mary Midgely was one such philosopher. Although not a Christian, she was highly critical of scientists' attempts to suggest that their discipline was a means for gaining intellectual 'salvation'.[1] (Perhaps due to her gender, she drew a rather less gentlemanly response from Dawkins than he had evinced in his discussion with Rowan Williams.) There are so many chapters in the development of STS and its critique of science that there is no space to describe them all here. One of the earliest and most famous examples is Thomas Kuhn's notion of 'paradigms' (1962), shared ways of thinking that define what counts as a legitimate scientific question and as a legitimate method for finding the answer. The radical implication of his theory is that there is no way of proving that one paradigm, one way of knowing, is better than another. They ask different questions and they have different standards of proof. Since Kuhn's time, many of the details of paradigms have

Science, Religion, and the Human Future. Amanda Rees et al., Oxford University Press. © Amanda Rees, Franziska E. Kohlt, Tom McLeish, Charlotte Sleigh, David Wilkinson (2025).
DOI: 10.1093/9780191995316.003.0012

been pulled apart and altered, but his basic insight remains in place. Subsequent researchers have demonstrated something like paradigms at work. For example, sexist assumptions have been baked into designs for technology, with male bodies and models used to design tools and machines that do not accommodate women. For many years, car crash dummies were based on the male form, resulting in higher injury and fatality rates amongst women. Racist assumptions have been baked into models of psychology, as we saw in Chapter 7; intelligence tests based upon European cultural norms confirmed white people as superior to the global majority.[2] Climate change science that shrinks the problem to a gas, rather than the behaviour of a culture, shrinks the solution to as-yet unproven machines for sucking that gas out of the air.

If it turns out that there are many paradigms of science, none demonstrably better than the other, then perhaps none of them is demonstrably better than faith, either. If we follow STS, are we back to granting that six days of creation is as good a theory as any other?[3] This fear sparked another 'war' that somewhat resembles the new atheist debates that followed a decade or so later. Last seen in Chapter 3, the 'science wars' were an academic spat in which a few scientists decided that they were being undermined by these critical perspectives.[4] It is perhaps fair to say that the emphasis of early STS was on critiquing science, highlighting its unsavoury dimensions. But as the seriousness of climate change became evident, their orientation underwent a subtle and profound change. As climate sceptics struck out at science, so STS scholars considered how they could ally themselves with scientists—yet without compromising their position that science can and should be done better.[5]

One of the best recent accounts of how science works comes in the opening chapter of *Why Trust Science?* (2019) by Naomi Oreskes. She begins by running through some of the proponents for scientific certainty, from the nineteenth to the twentieth century, and the reasons why their claims do not stand up. Oreskes then turns to the valuable contributions that feminist theory has made to scientific epistemology (ways of knowing) in recent decades. During the 1980s, feminist scholars began to critique scientism. For one thing, they pointed out how the value of objectivity could be understood as the intensive cultivation of European masculinity, a denial of the embodied self that went hand in hand with callousness towards the embodied subject—meaning: women, colonial subjects, and animals that lacked the capacity to do objective science. Carolyn Merchant explored the history of this trend, in an explosive book (1980) that highlighted the sexual violence present in Francis Bacon's descriptions of nature.[6] Bacon, often described as the father of experimental science, now seemed a most regrettable ancestor, the progenitor of cruelty and exploration in our treatment of the natural world.

Meanwhile, Sandra Harding and Helen Longino addressed similar themes from a philosophical perspective. Harding helped to develop the standard known as 'strong objectivity', an alternative to the supposed value-neutrality of science. Strong objectivity takes into account the perspectives of different people on a given topic. Similarly, Longino's feminist epistemology took Merchant's insights about gender and turned them into an all-purpose tool for noticing how human beliefs and values of all different kinds enable scientists to bridge the gap between evidence and certainty. Her conclusion was the same as Harding's: our science is more secure if we are aware of the places where our values play the part of invisible filler; the way to become aware of those values is to broaden participation in the scientific enterprise as widely and democratically as possible. Their philosophy has grown into the desideratum that is sometimes known as 'socially robust knowledge'. A 'solution' for climate change, for example, is no solution at all unless it has buy-in; all parties must see that it is equitable and achievable if it is to work. Vaccination doesn't prevent an epidemic unless everyone trusts it.

There are other ways that we could tell this story, other routes we could take through the development of STS towards new ways of thinking about science. If we told it through the work of historians, we might point to the work of Simon Schaffer and Steven Shapin, with their emphasis on how science must answer to contemporary social needs. If we told it through sociology, we could point to Harry Collins and his demonstration that it is not enough to trust a machine's measurement, but that we also have to trust the people who run the machines. If we told it through philosophical anthropology, we could point to Bruno Latour and his description of how machines, creatures, and people have to be assembled in networks of reliable behaviour.[7]

The point, however, is this: amongst those who believe there is a conflict between science and faith, it is usually presumed that faith needs to 'up its game' in order to meet the higher epistemological standards of science. If they read STS, they will discover that thinkers about science have made the opposite move for them, questioning the assumptions that underlie science's claim to have unimpeachable access to the truth. Science is revealed, through this work, to be an activity that is as human as any other, faith included. It builds remarkable achievements but does so through the human channels of consensus-making. Technological frameworks of instrumentation and measurement interpenetrate with systems of education, funding, and citation to produce conclusions that we, collectively, can lean upon. For Oreskes, the culmination of the story told via Harding and Longino comes with the Intergovernmental Panel on Climate Change (IPCC), whose work represents knowledge based in unprecedented scientific consensus. The first of the four reports that made up the Sixth Assessment (2021) was written by 234 scientists.[8] Each one of those scientists had written a PhD judged passworthy by an expert from outside their own university, according to the standards of their respective disciplines: botany, physics, chemistry, and so on. Each one used equipment that was tested and maintained to be reliable in different locations and conditions. Between them, they read over 14,000 scientific papers, each one the result of a study done up a mountain, from a satellite, in a lab. Each one of those papers was written by multiple authors and anonymously reviewed by experts. The authors and their reviewers had themselves gone through the process of training and vindication in their research methods and places of work. It was a vast spiders' web of interconnecting and self-critiquing activity: the biggest consensus the world has ever seen.

Going global

Around the turn of the twenty-first century, STS began to experience an energising influx of scholarship emerging in Postcolonial and Indigenous studies. This scholarship widened the critical demands of STS beyond feminism. Some studies highlighted the contributions of Indigenous and colonial subjects to scientific projects. Others argued that global majority cultures had 'beaten' the West to key scientific developments. Both of these narratives preserved the essential thrust of conventional, progressive histories. The most radical studies began to examine the nature and value of Indigenous Knowledge (IK) systems, neither as contributors nor antecedents to Western science, but as knowledge systems in their own right. Many of these refused the binaries that plague Western science, and do not separate the knowing subject from the world in which he or she lives.

In a recent book, for example, Lai Pak Wah emphasises how traditional Chinese medicine cannot be tested for its compatibility with Western medicine because of its different philosophical footing: *Wuxing* philosophy emphasises the relationships between things, whereas the experimental method is concerned with establishing the things themselves.[9] There have been many studies of Indigenous medicine and ethnobotany; typically, these communities do not treat plants or their products as extractable goods that can be used in any circumstance. Their appropriacy and efficacy are inextricable from the relationship and the circumstance in which they are administered.

Helen Verran has summarised some key studies (or stories, as she calls them) regarding Indigenous Knowledge.[10] One story concerns the Innu people of *Metsheshu Shipu* (Eagle River, Canada). Elders of the Innu Nation speak of '*Kanipinikassikueu*, the master of *atîku*'. *Atîku* seem to be what the settler-Canadians and scientists call caribou, but when the herd population collapsed in 2013, it became apparent that direct translation did not work at all. The elders refused to think of the *atîku* in isolation from *Kanipinikas-sikueu's* willingness to bless them. In this, they are different to scientists who, although they always measure length with a ruler (or intelligence with a test), believe that these qualities exist independent of their intervention in the world.[11]

Another story reported by Verran shows overlap between Indigenous and scientific thinking, combatting the belief that '[w]hereas modern science is rational, other knowledges ... are merely expressions of beliefs espoused by social groups'. The example concerns maps, which in this example are shown to capture collective stories, even in Western contexts of production. The story starts in 1770, when the master Indigenous navigator, Tupaia, created a hybrid map to translate between Captain Cook's time-based measurement of progress and his own method involving narrative and star-plotting. The fact that a hybrid model was possible suggested translatable components of narrative even in the 'scientific' model.[12]

Even mathematics, apparently the pure, distilled logic of science, is open to cultural variety, as a story from Nigeria reveals. Describing an inspired teacher who allows his class to use local forms of number, Verran reports that 'Yorùbá number is quite different in form to the numbers of science'. Although the two forms of numbering behave differently in use, their 'valuations of things like length tend to agree'. When both forms are connected in practice, Verran reports, 'a number that is neither and both a Yorùbá number and a science number comes to life'.[13]

With the exception of the history of science, most branches of STS have been reticent when it comes to considering the entanglements of science with the more familiar world religions such as Islam and Christianity. Hinduism is a growing area of interest; India combines technological ambition with the retention of Hindu icons and the enthusiastic practice of religious rituals, no less amongst scientists than amongst any other profession.[14] Nationalism frequently acts as a powerful and troubling conjunction of science and religion, with stories of India's traditional Vedic science used as a banner for Hindu scientific supremacy in the present.[15] However, scholars emphasise that the Indian conjunction of science and faith is more interesting than nationalistic propaganda might suggest, particularly when it comes to areas like transhumanism. Its rich compilation of technological and mythological beings, according to Banu Subramaniam, offers a 'generative possibility of myth and story ... that animates alternative imaginaries and "other" worlds of possibilities'.[16]

Subramaniam's focus upon avatars gives a new twist to science and religion studies not just in India but in many contexts. Avatars, originally the incarnations of Hindu gods, became associated with personhood and identity in games and online in the late twentieth century. As such, they indicate an important trend in how people think about science and religion. For all the twentieth-century fuss about the scientific method, many people today would see the key issue not as science but rather technology. Contemporary interest lies less in method and more in product. Scientists themselves are contemplating the possibility of using data to generate conclusions that do not arise from human hypothesis, but directly from patterns impossible for us to perceive. As we saw in the chapter on AI, we frequently relate to technology as if it were sentient, perhaps even a deity. AI will see into my problems. AI will fix the world. Treating AI as a god is sometimes done frivolously, and sometimes with deadly seriousness. Donna Haraway perceived the trend way back in 1985: 'Our machines are disturbingly lively, and we ourselves frighteningly inert'.[17] The coming decades are likely to see a recentring of 'science and religion' discussions towards a framework of 'science and technology'.

In the meantime, having come via IK, Hinduism, and other religions, STS, at last, shows signs of engaging with Christianity and other major religions. Christianity has served some time in academic exile for its indisputable crimes, including its collusion with colonialism. Academics of the Global North have begun to perceive that these problems afflict secular disciplines too, and to re-evaluate spiritually inclusive knowledge systems (usually in Indigenous and Postcolonial contexts) as sitting alongside, not beneath, Western science. Christianity may be making its return, in appropriately chastened form, as one of these. Contributors to the volume *Science and Religion: Approaches from Science and Technology Studies* (2024) show how productive this conjunction can be.[18] They highlight how STS enables one to take a non-essentialist approach to both science and religion, focusing more on practice rather than the ultimately unprovable questions of belief. STS does not pretend to be neutral or objective in the way that science generally does. Implicitly or explicitly, its critique is often normative, meaning that it points out when voices are erased or excluded in knowledge-making. Its underlying mission is often transformative: how can science and technology be made more equitable for the human and more-than-human world?

Tell me the old, old story

As we mentioned in passing, Verran prefers to think of her examples not as case studies but as stories. Her reasoning for this choice springs from her allyship with the IK of which she writes:

> Knowledge authorities in Indigenous communities are often said to tell stories, whereas scientists speaking authoritatively from state institutions are said to wield theories. However, if we examine the difference between story and theory through the lens of the words themselves, and consider the difference in terms of the stories the words themselves carry, we see that a 'histōr' and a 'theor' in Ancient Greece were differentiated only in terms of power. A histōr spoke authoritatively according to those of his own time and place; a theor carried the message of an authority approved of by those who governed.[19]

In keeping with her commitment to questioning the governing force of Western science, Verran chooses to call her own work 'story'. A story does not belong to any one person but rather emerges from a covenantal relationship between teller and listener. If there is no listener, there is no story. A story can be passed on, but it is only fully alive in a community that knows how to share it, when to share it, and what questions to ask of it.

In this book, we have seen many stories shared, whether biblical or scientific. We have seen how they can create harmonious and fractious relations between science and religion; indeed, they have, in some cases, *produced* those categories. Stories are fundamental to the ways in which most people organise their worldview. They are also risky because we often think *with* them, rather than *about* them.

Cultural scholars and historians have been debating and analysing the role that narratives play in public culture for some time. They have a long history in many Christian traditions, where personal testimony, offered in public, can be a prompt for worship or a call to the 'unsaved'. Carlo Ginzberg and Lawrence Stone have closely analysed the role that narratives—stories— play in the telling of history.[20] Hayden White argued that thinking in narratives 'is so natural to human consciousness' that challenging the way they structure our world requires our close attention and analysis.[21] More- over, because narrative simultaneously describes and explains the world, we can never get behind the story to the truth. No narrative account of event can be taken as an unproblematic recounting of the facts of the mat- ter. Nor is it even as simple as accepting that each narrator inevitably tells their story from their own point of view. Instead, White argued that narrative itself *moralises*, or makes the story meaningful: that 'narrativity is intimately related to, if not a function of, the impulse to moralise reality'.[22] White's med- itation was initially related to his efforts to differentiate more clearly between history and literature on the one hand, and history and natural science on the other, arguing that history's use of narrative rendered history's status as 'science' suspect. However, as later scholars have shown in studying popu- lar science and the public understanding of science, the use of narrative in science communication is ubiquitous—and equally challenging.

There is a well-established history of science communication that describes in rather unkind terms the efforts of the 1980s. At this time, as we have seen in earlier chapters, British and American scientists felt under attack, with the nuclear bomb and pesticides unfairly taken as emblematic of their work. The problem, as a famous study commissioned by the Royal Society concluded, was that the public did not know enough about science. If they knew more, they would love it more. Therefore, the solution was to open their brains and pour in as much science as possible. STS scholars and others dubbed it 'the deficit model'.

Critics were swift to point out that lecturing people is not the way to enthuse them; there was no reason to believe that knowledge about nuclear science, for example, would wipe away legitimate anxieties about its development. It might, on the contrary, lead to more and better-informed criticism. Thus, the model of 'dialogue' was proposed as a preferable alter- native, employed by the British government in relation to the GM debates of the 1990s (Chapter 5). Dialogue encouraged the public to bring their own stories to the table, the kinds of stories rooted in experience that Brian Wynne had found amongst the Cumbrian sheep farmers. The need for pol- icymakers, as well as scientists, to take stories seriously has been eloquently made by Sarah Dillon and Claire Craig, who also emphasise the collective nature of stories. It is not enough simply for scientists to tell a story. *Listening* to the kind of stories that are being told about controversial science and understanding how these stories fit within wider context of community are vital.[23]

Over recent years, marketers and media leaders have also come to emphasise more and more the power of story, especially autobiography. This was true in the field of television and has become even more so with the rise of social media and the influencers who brought their lives (or a narrative thereof) to the screens in everyone's hands. The concept of relatability skyrocketed with the introduction of smartphones, indicative of the belief that social media narratives told universally recognisable personal stories—and as such held opportunity for profit.[24] Dialogue proved to be difficult to achieve in conversations about science, and was illusory in the echo-chamber of advertising that Internet 2.0 turned out to be.

In the United Kingdom, the Wellcome Collection has led the way in creating stories about science. The Wellcome Trust has a focus on medical science, and the Collection's exhibitions and online 'Stories' have emphasised personal experience as a way of understanding and humanising topics of health, dis/ability, bodies, and disease. Other recent projects include the LSE initiative Narrative Science, which roots itself in STS scholarship. Unsettling Scientific Stories, an Arts and Humanities Research Council funded project based at the University of York, focused on science narratives directed towards the future, while in news recent at the time of writing, the British government convened a group of science fiction writers to help it think about future technological threats.[25]

The kind of science that is likely to appear in the public domain is relatively unusual. Basic science is unlikely to be covered, unless some innovation has upended previous understandings of the way the world works. New discoveries such as the identification of a new planet, or the development of novel technologies, or the announcement of the winners of the Nobel prizes, are much more likely to be found in the science sections of the media. At the same time, science in the media also often focuses on subjects that its audiences find relevant, which is why so many examples of science in the media revolve around matters of health. Other media-friendly science includes stories about origins, whether of the universe or of humanity, or the activities of charismatic megafauna, alongside accounts of how new technologies *may* directly change human lives and expectations. What unifies this diverse range of subjects is the narrative structure by which they are told. They open with an exposition of the situation or problem that is being faced (beginning), before delving into a range of possible solutions to that problem (middle), until the problem is resolved (end), with perhaps a nod to further questions that will need to be considered. Two examples from the newspaper on the day of writing make this point.[26] One concerns charismatic megafauna, cockatoos, which other scientists had noticed dunking rusks into water before eating them. A range of crudités and dips were presented to the birds, yielding the conclusion that pasta and blueberry soy yogurt was the favoured and 'innovative' combination. The question of whether the behaviour was innate or learned was flagged as a coda. A second story reverses the conventional order, beginning with the conclusion, a 'lucky' photograph of gravity bending light that proves the long-ago predictions of Einstein (the problem). The story is presented as evidence of the success of the new Euclid telescope, and the mysteries of dark matter are left hanging for future resolution.

Using this structure doesn't just make sense from the point of view of storytelling. It also mirrors quite closely the version of the scientific method that is often described or implied in public. This model also involves the identification of a problem and an account of previous attempts to solve said problem (beginning stage). Various hypotheses or solutions are proposed and tested or falsified by the gathering of evidence (middle stage). Finally, there is a tentative embrace of the hypothesis or solution (end). This account represents an *idealised* version of scientific practice. As the Nobel-prize–winning British biologist and writer Peter Medawar famously suggested in 1963, it describes the structure of a research paper, rather than what scientists actually *do*.[27] Medawar provocatively entitled his talk 'Is the scientific paper a fraud?'. He was particularly keen to highlight the way in which the structure of published research underplays the role of inspiration and creativity in scientific research.

In the years that followed Medawar's talk (though following other prompts), ethnographers and historians of science followed scientists into the lab and field to see what they actually did. They too found a serious mismatch with the accounts that appeared in peer-reviewed scientific papers. Science in practice was much messier than the stories enshrined in accounts of 'the scientific method'. In reality, the findings of scientists were impacted by the skill, experience, and resources of the experimenter, as well as beliefs about which facts should 'count' when assessing the success of an experiment (c.f. Longino).[28] None of this, however, detracted from the importance of the idealised version of science that lay at the heart of the account of 'the scientific method', consistently invoked to differentiate science from every other form of human activity.

The sociologist Ron Curtis has pointed out that popular science writing also shows a close relationship with another form of narrative: the detective story. This is significant because, while this specific genre is ostensibly oriented to problem-solving, its narrative structure simultaneously imposes limitations on the capacity of the audience to challenge the account of the world that is being presented. Detective stories, Curtis showed, similarly begin with a problem (the crime) which is followed by the patient gathering of facts, either by the detective (Arthur Conan Doyle) or by skilful interrogation of those who, knowingly or not, already possess the facts (Agatha Christie). Rarely, a story might acknowledge the routine inquiries carried out by the detective's colleagues or underlings that characterises real police work. Likewise, it is rare to see the routine activities of technicians and students given much space in popular accounts of science. Having gathered the relevant information, the hero-detective is able, often by a process of elimination, to identify the criminal. Their culpability is not in doubt, framed by the narrative. As Curtis puts it, both detective stories and popular science narratives might 'begin with unanswered questions', but we always 'end with unquestioned answers'.[29] Popular science, written according to a narrative structure, is a powerful instrument for maintaining the cultural authority of science.

The fact that popular science stories emphasise the inevitable success of the scientific method has the unfortunate side effect of setting scientists up for failure in the context of controversy or crisis. In the chapters of Section II (Covid-19, genetic modification, climate change), we saw examples of ordinary scientific uncertainty being interpreted as a catastrophic failure of science. If people are led to believe that science will give them clear answers, then any failure or even hesitation will understandably lead to suspicion and confusion. It is ironic and frustrating that in this context, scientists are sometimes forced into playing exactly the same role as that which the conflict thesis attributes to faith leaders: dispensers of truth and unquestioned guides on the road to salvation.

Towards relationship

In this chapter, we have seen some large trends that have shifted the discourse away from conflict and towards cautious collaboration in the making of knowledge that is meaningful and useful for all. These prominently include the efforts of scholars outside both science and theology, in the field of STS. Meanwhile, activism in Postcolonial and Indigenous settings has challenged dominant view of science as all-powerful. This, in turn, has further informed global scholarship on science and religion, clearing space for a rediscovery of the value of Christianity and other major religions in thinking about science and technology.

Public discourse takes time to catch up with academia. It requires good storytelling, as we have also discussed within this chapter. Stories on their own, however, are not enough. Good stories require good relationships between teller and listener, a connection built on trust: not the fake trust demanded by the 'relatable' influencer, but something authentic. Part of a trusting relationship is an expectation for the roles to reverse. Let me listen to your story, and I will tell you mine. We have seen how science communication fell at the hurdle of the deficit model, attempting to talk at people rather than listening to them. The dialogue model proved difficult to implement in practice, but steps have been taken via citizen science (the inclusion of ordinary people in research projects) and citizens' assemblies.[30] Citizens' assemblies convene a representative cross section of the public to hear experts speak on a specific topic before offering their decisions on how a scientific or technological problem should be addressed. The United Kingdom made use of this model in 2020 to address climate change policy, although there has been criticism about the model's tendency towards blandness—and, more significantly, the government's failure to respect its outcomes. This last is again a failure of trust.

Faltering though these steps towards trust may be, they are perhaps the best model we have for respectful discussions in science and faith. Whether in policy settings or in congregations, hearing the testimonies of science from all angles helps us to build collaborative, multifaceted stories. This book's conclusion describes experiments conducted along these lines.

Throughout this book, we have looked under the lid of overly simple stories about science and faith. We have seen how the story about good science replacing bad religion does not hold water as a description of history. We have also seen how the creation of that story was an attempt on the part of some scientists to create power and authority for themselves: to demand the trust of a public increasingly ruled by 'rationality' and technology. At the same time, we have seen how some of the damaging and destructive stories about and within science have borrowed their script from stories about and within Christianity. The civilising mission, or the coming of progress, was our biggest example, justifying exploitation and violence in the interchangeable names of capital and Providence.

We have also seen theological stories tangled up in the development of new crops, vaccines, and AI. These specific sciences, and others like them, are a good place to begin. The approach is, like Polkinghorne's, bottom-up; in acknowledgement of the contributions of STS, it is also self-critical. With the refreshed relevance of faith and spirituality that this and the previous chapter have described, there is a hunger to think about their theological dimensions.

Further reading

Harry Collins, Robert Evans, Martin Innes, Eric B. Kennedy, Will Mason-Wilkes, and John McLevey, *The Face-to-Face Principle: Science, Trust, Democracy and the Internet* (Cardiff: Cardiff University Press, 2022)

Zara Thokozani Kamwendo, ed., *Science and Religion: Approaches from Science and Technology Studies* (London: Palgrave Macmillan, 2024)

Patty Krawec, *Becoming Kin: An Indigenous Call to Unforgetting the Past and Reimagining Our Future* (Minneapolis: Broadleaf Books, 2022)

Will Mason-Wilkes and Alexander Hall, *Most Adaptable to Change: Evolution and Religion in Global Popular Media* (Pittsburgh: University of Pittsburgh Press, 2024)

Mary Midgely, *Science and Salvation: A Modern Myth and Its Meaning* (London: Routledge, 1992)

Notes

1. Mary Midgely, *Science and Salvation: A Modern Myth and Its Meaning* (London: Routledge, 1992).
2. Ruha Benjamin, *Race after Technology: Abolitionist Tools for the New Jim Code* (Cambridge: Polity Press, 2019); Caroline Criado Perez, *Invisible Women: Data Bias in a World Designed for Me*n (London: Vintage, 2019).
3. Steve Fuller, 'Intelligent Design Theory: A Site for Contemporary Sociology of Knowledge', *Canadian Journal of Sociology/Cahiers canadiens de sociologie* 31.3 (2006): 277–289, https://doi.org/10.2307/20058711.
4. Michael Lynch, 'We Have Never Been Anti-Science: Reflections on Science Wars and Post-Truth', *Engaging Science, Technology, and Society* 6 (2020): 49–57, https://doi.org/10.17351/ests2020.309.
5. Bruno Latour, 'Why Has Critique Run Out of Steam? From Matters of Fact to Matters of Concern', *Critical Inquiry* 30.2 (2004): 225–248, https://www.journals.uchicago.edu/doi/epdf/10.1086/421123.
6. Carolyn Merchant, *The Death of Nature: Women, Ecology, and the Scientific Revolution*, 40th Anniversary edn. (New York: HarperCollins, 2019).
7. Steven Shapin and Simon Schaffer, *Leviathan and the Air-Pump: Hobbes, Boyle, and the Experimental Life* (Princeton: Princeton University Press, 2011); Steven Shapin, *Never Pure: Historical Studies of Science as if It Was Produced by People with Bodies, Situated in Time, Space, Culture, and Society, and Struggling for Credibility and Authority* (Baltimore: Johns Hopkins University Press, 2010); Bruno Latour and Steve Woolgar, *Laboratory Life: The Construction of Scientific Facts* (Princeton: Princeton University Press, 2013);

Harry Collins, *Changing Order: Replication and Induction in Scientific Practice* (Chicago: University of Chicago Press, 1992).

8.   Stephanie Spera, '234 Scientists Read 14,000+ Research Papers to Write the IPCC Climate Report—Here's What You Need to Know and Why It's a Big Deal', The Conversation, 9 August 2021, https://theconversation.com/234-scientists-read-14-000-research-papers-to-write-the-ipcc-climate-report-heres-what-you-need-to-know-and-why-its-a-big-deal-165587, accessed 11 February 2025.

9.   Lai Pak Wah, *The Dao of Healing: Christian Perspectives on Chinese Medicine* (Singapore: Graceworks, 2018).

10. Helen Verran, 'Indigenous Knowledge Traditions and Science', in *Elgar Encyclopedia of Science and Technology Studies*, edited by Ulrike Felt and Alan Irwin (Cheltenham, UK: Edward Elgar Publishing, 2024), https://doi.org/10.4337/9781800377998.ch29. It is regrettable that the following paragraphs adopt the Western practice of extracting these stories from their original communities and reproducing them as tradable points in an argument. The reader is encouraged to participate in their own stories.

11. Helen Verran, 'Indigenous Knowledge Traditions and Science', in *Elgar Encyclopedia of Science and Technology Studies*, edited by Ulrike Felt and Alan Irwin (Cheltenham, UK: Edward Elgar Publishing, 2024), https://doi.org/10.4337/9781800377998.ch29, p. 285.

12. Helen Verran, 'Indigenous Knowledge Traditions and Science', in *Elgar Encyclopedia of Science and Technology Studies*, edited by Ulrike Felt and Alan Irwin (Cheltenham, UK: Edward Elgar Publishing, 2024), https://doi.org/10.4337/9781800377998.ch29, p. 287.

13. Helen Verran, 'Indigenous Knowledge Traditions and Science', in *Elgar Encyclopedia of Science and Technology Studies*, edited by Ulrike Felt and Alan Irwin (Cheltenham, UK: Edward Elgar Publishing, 2024), https://doi.org/10.4337/9781800377998.ch29, p. 288.

14. Robert M. Geraci, *Temples of Modernity: Nationalism, Hinduism, and Transhumanism in South Indian Science* (Lanham, Maryland: Rowman & Littlefield, 2018).

15. Meera Nanda, 'Science Sanskritized: How Modern Science Became a Handmaiden of Hindu Nationalism', in *Routledge Handbook of South Asian Religions*, edited by Knut A. Jacobsen (London: Routledge, 2020): 264–286.

16. Banu Subramaniam, *Holy Science: The Biopolitics of Hindu Nationalism* (Seattle: University of Washington Press, 2019).

17. Donna Haraway, 'A Cyborg Manifesto: Science, Technology, and Socialist Feminism in the Late Twentieth Century', in *Simians, Cyborgs and Women: The Reinvention of Nature* (New York; Routledge, 1991): 149–181; p. 152.

18. Zara Thokozani Kamwendo, ed., *Science and Religion: Approaches from Science and Technology Studies* (London: Palgrave Macmillan, 2024).

19. Helen Verran, 'Indigenous Knowledge Traditions and Science', in *Elgar Encyclopedia of Science and Technology Studies*, edited by Ulrike Felt and Alan Irwin (Cheltenham, UK: Edward Elgar Publishing, 2024), https://doi.org/10.4337/9781800377998.ch29, p. 284.

20. Elizabeth Clark, *History, Theory, Text: Historians and the Linguistic Turn* (Harvard: Harvard University Press, 2004).

21. Hayden White, 'The Question of Narrative in Contemporary Historical Theory', *History and Theory* 23.1 (1984): 1–33, https://doi.org/10.2307/2504969, p. 1.

22. Hayden White, 'The Value of Narrativity in the Representation of Reality', in *On Narrative*, edited by W. J. Mitchell (Chicago: University of Chicago Press, 1981): 1–24, p. 14.

23. Sarah Dillon and Claire Craig, *Storylistening: Narrative Evidence and Public Reasoning* (London: Routledge, 2021).

24. https://books.google.com/ngrams/graph?content=relatable&year_start=1800&year_end=2022&corpus=en&smoothing=3, accessed 11 February 2025.

25. Amanda Rees and Iwan Rhys Morus, eds., *Presenting Futures Past: Science Fiction and the History of Science, Osiris* 34.1 (2019), https://www.journals.uchicago.edu/toc/osiris/2019/34/1; https://www.narrative-science.org/ accessed 11 February 2025; https://www.theguardian.com/uk-news/2025/jan/19/ministry-of-defence-enlists-sci-fi-writers-to-prepare-for-dystopian-futures, accessed 18 February 2025.

26. https://www.theguardian.com/environment/2025/feb/10/cockatoos-show-appetite-for-dips-when-eating-bland-food-find-scientists, accessed 18 February 2025; https://www.theguardian.com/science/2025/feb/10/euclid-telescope-captures-einstein-ring-revealing-warping-of-space, accessed 18 February 2025.

27. Peter Medawar, 'Is the Scientific Paper a Fraud?', *The Listener* 70 (1963): 377–378.

28. Bruno Latour and Steve Woolgar, *Laboratory Life: The Construction of Scientific Facts* (Princeton: Princeton University Press, 2013); Harry Collins, *Changing Order: Replication and Induction in Scientific Practice* (Chicago: University of Chicago Press, 1992).

29. Ron Curtis, 'Narrative Form and Normative Force: Baconian Story-Telling in Popular Science', *Social Studies of Science* 24.3 (1994): 419–461, https://doi.org/10.1177/030631279402400301.

30. Massimiano Bucchi and Brian Trench, eds., *Routledge Handbook of Public Communication of Science and Technology* (Abingdon: Routledge, 2021).

Conclusion

Everyone loves a good fight

As this book was first drafted in 2023, two blockbuster movies dominated popular culture: Christopher Nolan's *Oppenheimer* and Greta Gerwig's *Barbie*. They had a simultaneous release date, leading to some double-bill cinema attendance, sometimes in costume, known as 'Barbenheimer'. The dual release was playfully hyped by the media as a conflict in itself: who would win the battle of the box office? In the event, *Barbie* was number one and *Oppenheimer* number three in the list of highest-grossing movies of 2023.[1]

At one level, a fantasy comedy about a fashion toy and an epic biography about a physicist and pioneer of the atomic bomb have little in common; this was indeed the joke behind Barbenheimer. Yet both share a central narrative of conflict. *Oppenheimer*, based on the 2005 biography *American Prometheus*,[2] tells the history of the development of the Manhattan Project through a lens of conflict between Robert Oppenheimer and the United States Atomic Energy Commission Chairman Lewis Strauss. It presents Strauss as pursuing a vendetta against Oppenheimer for having humiliated him, publicly mocking his concerns about exporting radioisotopes and privately putting him down in a conversation with Einstein. In retaliation, Strauss sets up a private hearing revoking Oppenheimer's security clearance, thus limiting his influence in public and political discussions of nuclear policy.

The clash of these two people, the scientist and the politician, the establishment outsider and insider, is a powerful way of telling a complex story. Its power, however, is also its weakness. Oppenheimer's relationship with other members of the scientific community is underplayed (particularly with Edward Teller). Perhaps more importantly, the complexity of Oppenheimer's inner struggles is not fully represented. As we explored in earlier chapters, science often presents the pursuit of research as separate from its

Science, Religion, and the Human Future. Amanda Rees et al., Oxford University Press.
© Amanda Rees, Franziska E. Kohlt, Tom McLeish, Charlotte Sleigh, David Wilkinson (2025).
DOI: 10.1093/9780191995316.003.0013

application; if ever there was a story that gives the lie to such a guiding myth, this is it. The history of how cutting-edge research is done by individual scientists both within overlapping structures of scientific community and politics is far more interesting and subtle than even a three-and-a-half-hour movie could explore.[3]

The story of *Barbie*'s eponymous hero and her search to find her self-identity exposed another type of conflict. The film addressed and to some degree subverted the sexism and capitalism which have led to deep conflicts in Western societies. Helen Mirren's opening narration concludes tongue in cheek: 'Thanks to Barbie all problems of feminism and equal rights have been solved', and from there the movie explores questions of femininity and masculinity, conflict, and community. The embodied conflict of the Barbies and the Kens provides a fictional narrative device to explore real and deep-seated conflict and injustice. A number of journalists placed *Barbie* alongside the *Eras Tour* of Taylor Swift as an exploration and critique of patriarchy.[4] Michelle Goldberg commented that 'beneath their slick, exuberant pop surfaces, [both the film and the tour] tell female coming-of-age stories marked by existential crises and bitter confrontations with sexism'.[5] The populist right has upped the ante on conflict even further, grounding it firmly within the so-called culture wars over ethnicity, race, and sexuality, as well as gender. At the time of revising this book (2025), the new US regime was eager to get its retaliation in first: officials were mandated to withdraw funding from research projects that described themselves with reference to newly banned words such as 'female', 'disability', 'LGBT', or 'black and latinx'.[6] Equity and diversity initiatives in public life were actively targeted. There is no space for dialogue, it seems, only for fighting.

There are many voices within traditional and new media who buy not only into the conflict myth but also into the additional myth that 'conflicts sells'. One of this book's authors, David Wilkinson, recalls tackling the producer about an episode of the television programme *Newsnight*, in which a vocal atheist and equally pugnacious six-day creationist 'discussed' their differences in a conversation that was 'as constructive as mud-slinging'. The producer responded, simply, that conflict is engaging. Another BBC flagship programme, the quiz show *University Challenge* once asked the question: 'which scientist was opposed by the church for using chloroform anaesthesia for women giving birth?' The answer given, James Young Simpson, was judged correct, but the question itself was wrong. As Colin Russell has shown, the evidence for this resistance is almost non-existent.[7]

In the past decade, social media has surpassed television in terms of its influence, but there has been little sign of an increase in the diversity of voices that consumers encounter. The algorithms of social media have rather provided echo chambers to strengthen specific conflict narratives.[8] While the BBC may have learnt its lesson on balance in the presentation of climate change, climate deniers have found new outlets. A 2022 study of Twitter (currently X), Facebook, and Instagram found not only that hashtags such as #climatescam were rising dramatically in use but also that ads were increasingly being placed by fossil fuel supporters.[9] In short, the report found 'a stark comeback for climate denial'. Another recent study tracked the digital footprints of climate scientists and deniers across multiple climate change publications. The study showed that about half the mainstream media visibility still goes to climate-change deniers and that this proportion increases significantly when blogs and social media outlets are taken into account.[10]

Jeremy Shapiro calls attention to the dichotomous, all-or-none thinking that continues to afflict conventional and social media.[11] He argues that such black-and-white thinking is a source of dysfunction in mental health, relationships, and politics. It simplifies the world into binaries, rather than a space that offers many possibilities. When it comes to science, we see attempts to present information in a way that is relevant to specific audiences, to organize central ideas to resonate with their core values, and to pare down complex issues.[12] Conflict narratives have been used, and continue in use, to frame the creation of new institutions and identities.

The alliance of science

The thirst for conflict is nothing new. The preceding chapters have included many examples. The Galileo affair involved genuine conflict between Galileo and the church as shown in trials, recanting, and house arrest. The clash of imperial science with indigenous cultures brought conflict in the shape of injustice and violence. The nineteenth-century emergence of institutions of science within universities and other autonomous organisations provoked conflict with those who wanted to maintain power and control via other means. One feature of this latter event was the sowing of conflict between the coalition of partners that originally gave birth to the civilising mission. Some scientists turned against faith; meanwhile, some of the missionaries who had been carried on the wave of trade and military conquest turned into opponents of the system that had made their travels possible.

And time after time in the twentieth century, scientists had moments of conflict with the agendas of politicians and big business in environment, public health, and genetic modification.

At the same time, however, this volume has echoed other scholarly studies in finding that Christian faith and science are by no means implicitly or necessarily in conflict. We echo the view of Colin Russell that '[t]he conflict thesis, at least in its simple form, is now widely perceived as a wholly inadequate intellectual framework within which to construct a sensible and realistic historiography of Western science'.[13] Our survey of medieval science demonstrated most powerfully the co-development of what are now seen as two different activities. We have further argued that the imposition of a conflict narrative of science and religion onto the telling of the historical story has itself masked, misrepresented, and mauled important lessons from moments of genuine conflict. We have seen that when the academic world collides with popular media culture, the complexity of history is often flattened into a simplistic polarising weapon. In the arena of cosmology, for example, Stephen Hawking's growing atheist theology was covered at the expense of scientific critiques on his formulation of quantum gravity and the variety of theological responses to his claims by fellow scientists.

Still more prominent is the widely told myth that Galileo was condemned by the church for contradicting the Bible. While there are a few documented remarks from preachers of the time, using the Bible to say that the Sun revolved around the Earth, these are rare. John Calvin argued against Galileo not from a biblical perspective but on the common-sense basis that one cannot feel the Earth moving. What was at stake in the Galileo story was something far more important: Galileo's decision to point his telescope at the skies. The Aristotelian universe, baptised by Thomas Aquinas in the theology of the church, embedded the Hellenic credo that science was done primarily by the application of philosophical logic. Galileo's emphasis on the primacy of observation, his method of doing science, was the real conflict. As Galileo's biographer has commented:

> It is a curious fact that historians have not blamed philosophers rather than theologians for the decision taken against freedom of scientific opinion in astronomy. Yet philosophers alone urged the intervention of theologians, confident they would be on their side.[14]

We can place Galileo in a continuous tradition with thinkers such as Robert Grosseteste (Chapter 2), deploying a Judaeo-Christian methodology of looking at the world to see what God had created rather than trying to work it out through human logic.

In yet another twist, this book has argued that a lack of conflict is not necessarily a good thing. Theology is sometimes bad, and when bad theology informs science, terrible things can happen. The eighteenth and nineteenth centuries, as described in Chapter 3, are story of distressing collusion between science and theology. Focused upon 'progress', they produced a joint message about a salvation through rationality that was mediated via technology and Providence. Its colonial outworkings continue to reverberate in the present day. US ambitions to colonise space are perhaps this book's most obvious example. We also saw that AI folds into itself imperialist and racist definitions of intelligence, and that church-and-state efforts to support vaccination during the recent pandemic were at risk of creating an irrational, non-white other. We have further argued, in perhaps this volume's most original aspect, that science always has an implicit theology and that this in turn requires critique. Efforts to hamper climate change have been hampered by a thin theology of science that mirrors a distorted evangelical doctrine of belief as means to salvation. The claim that scientists can separate their research from social context is perhaps the most powerful example of bad theology, a claim to a 'view from nowhere' that elides into a 'God's-eye view'. This was the key argument of the chapter on genetic modification, but could be applied to most of the examples in Section II. (Russian cosmism is one of the notable examples: a different theology of science.) It is a claim to objectivity that can only arise from the systematic suppression of empathy, both on the level of self-discipline and in an institutional framing that keeps fact and value apart.

If we accept that science has implicit theology, then the space for fruitful dialogue increases. Section III has highlighted opportunities for, examples of, and failures in such dialogue over the past couple of decades. We saw, in Chapter 9, examples of science communication that create saints out of biologists, physicists, and engineers. Rather than condemning such sanctification—an entirely natural feature of human behaviour—a theological approach would rather ask on what grounds we canonise. It would ask what features of moral and spiritual life are being upheld in our choices, and what behavioural effects they might have on readers and consumers of these myths.

We also saw, in Chapters 10 and 11, how a variety of people began to critique seriously the myth of the fact/value divide upon which modern science is premised, and of which the new atheists represented a late gasp.

These critics included scientist-theologians, STS researchers, and Indigenous activists and scholars. It has even been suggested that STS can be regarded as a type of theology in itself, practising in the liberationist mode.[15] The efforts of these scholars have limned out an interdisciplinary space where fruitful dialogue that respects the complexities and histories of science and faith can occur.

Creating shared futures

Life on this fragile planet demands, more than ever, great thought and insight concerning futures shared between different groups of people, and between people and the more-than-human world. A liveable future will be created not through conflict but through shared practices of imagination and compassion. It will entail reconnecting the fields of fact and value that have been driven apart into the reservations known as science and faith, respectively. Reconnection or re-ligation: the origin of the Latin term *religio*, according to St Augustine. Theology, deployed as thoughtful critique or re-reading of science (another possible origin for *religio*), is something that can be applied to science in order to remake the connection between fact and value: to create a science that works for shared futures.

This challenge lies at the heart of the work of the project Equipping Christian Leadership in an Age of Science (ECLAS).[16] Active since its pilot in 2013, ECLAS is funded by the John Templeton Foundation, based at the Universities of Durham and York, and is the project behind this volume. Through all its activities, ECLAS addresses one central question: 'How does context give challenges and opportunities for equipping Christian leadership in an age of science?' The research of sociologist Lydia Reid, exploring the specific question of how faith leaders think about science, has been a major impetus to its efforts.[17] Based on a major survey of over 1000 church leaders and in-depth interviews of over forty senior leaders, Reid showed that these people have a generally positive view of the relationship between science and Christian faith. However, this is coupled with a hesitancy when it comes to engaging in public discussions. One bishop commented, 'What I sense with quite a lot of our clergy ... is that they're rather frightened of the subject because of the Dawkins antagonism ... they feel they're treading in very dangerous territory—that they'll be shown up for not knowing enough about it, they think they're inexpert ... [they] begin to be even more frightened of science because they see it as something that's antagonistic to faith'.

Another Bishop lamented: 'It's like an acid rain wearing away the confidence of Christians'.

Such reluctance can have a ripple effect within the church and beyond. Silence on the part of senior leaders in the context of a culture that often takes the conflict narrative for granted can be read as lending credence to the division. (It can also have the more specific effect of undermining the vocation of scientists in church congregations.) Reid found considerable evidence that the how/why distinction of science/faith—Stephen Jay Gould's non-overlapping magisteria—can sometimes be used as an avoidance strategy for fear that scientific insights might challenge deeper-level understandings of how God works in the world.

There is, however, substantial evidence of faith-leaders engaging with science privately even as they refuse to speak of it in public. In a further set of in-depth interviews of twelve Bishops of the Church of England during the pandemic, the sociologist Thoko Kamwendo (another member of the ECLAS project) showed that their awareness of history and the conflict narrative helped frame their apprehension of the pandemic, often leading to a reluctance to interpret the pandemic theologically.[18] They were deeply aware of the fraught history of Christian commentators interpreting natural disasters as divine action. This made it difficult for them to create a theological space in which suffering caused by the pandemic, and the virus itself, could be understood, not as 'the enemy', but as part of creation. This also made it harder for leaders to maintain their focus on human responsibility and care for others in their response to the pandemic.

Conferences organised by ECLAS give safe space for Christian leaders to engage with scientists on various topics.[19] 'Safe space' might seem like an unnecessary offering to such powerful people, but when it comes to science, Reid's findings are precisely that the fear of making scientific mistakes, or looking foolish, do hold back their participation in public dialogue. The aim is not to build confidence amongst church leaders by teaching them science, but to allow them to experience the vocation, passion, and questions of fellow human beings who are scientists. Our themes have included AI, fracking, neuroscience, agriculture, extinction, cosmology, and many more. Those that have been conducted in person have included lab visits, so that science is encountered 'in the making' and not just as an apparently polished set of facts. Those conferences that have been online have benefited from contributors with situated knowledge of various kinds (c.f. Chapter 5). One conference on food and climate was memorable for its testimonies from participants at the frontline of crop precarity. Kenyan bishop Emily Awino

Onyango, for example, gave voice to a dimension of climate change that is not obvious to those in the Global North, namely, that gender violence is also exacerbated by the crisis. A publicly disseminated and interactive 'Climate Cathedral' resource was developed in connection with this conference, with emphasis on the co-creation of science and theology (that is, 'facts' and 'values') and the participation of situated experts.[20]

ECLAS is also speaking to the pipeline of future church leaders via its 'Science in Seminaries' project.[21] This scheme encourages and equips theological colleges to develop courses that normalise discussions about science early on in the process of formation for ministry. It is up to different colleges to decide what topics and themes will best complement their local culture. To date, these have leaned towards themes of ethics, health, and the environment, along with some interesting explorations of science in relation to specific denominational traditions in the Pentecostal and Baptist churches.

At a more grassroots level, ECLAS supports scientists in their local churches via 'Scientists in Congregations'. These projects start with an established relationship between a local church leader and a professional scientist, focusing on strategic churches that have strong ties with local communities. The scientist within the partnership is sometimes a member of the congregation and sometimes the colleague or friend of the church leader; it is generally someone who has already engaged in conversation around faith and science. Together, partnership generates a creative project that involves a local congregation and invites the public more broadly. Past projects have included art installations, science fairs, plays, workshops, and lecture series. As in the Leadership Conferences and Science for Seminaries strands, the themes are always scientifically specific.

Another ECLAS strand sits within the theme of storytelling highlighted in Chapter 11. The 'New Narratives' project aims to rewrite relations between science and faith via short narrative blogs about how the two have interacted.[22] There is little or no explicit mention of the conflict thesis, but a consistent and implicit rebuttal via its content. For readers accustomed to STS and science-engaged theology (SET), the fundamental message of interaction will not be new, but for many readers, it is hoped that it will be a surprising and refreshing reframing of the conflict. Stories already published include a nun who became an accomplished microbiologist in order to make better cheese; the role of parish records in historic epidemiological research; and the teachings of plants. The stories have been created by writers trained in STS and, as such, they challenge the myths of value-free science. They present as examples of good science specimens that are characterised by their engagement with situated expertise, such as peer-led perinatal care and Indigenous astronomy.

A final ECLAS strand relates to the unusual governmental composition of the UK, where Anglican bishops sit in the upper house (the House of Lords). Because of this, ECLAS has had a unique opportunity to develop the connections between science and theology at an active site of public policy-making. ECLAS co-director Kathryn Pritchard works within the Church of England's Faith and Public Life Department and has helped to foster spaces where senior scientists and church leaders can build relationships of trust on issues such as fracking and AI. The conflict thesis rarely manifests itself in these conversations, which tend rather towards the complex, human dimensions of science, and to questions of values and justice, such as previous chapters in this book have explored. Scientists have been impressed by the way that church leaders are involved in the concerns of their communities and by the insights produced through their traditions of theological reflection.[23] This unique British niche for ECLAS is, of course, a model that has limits for thinking about the rest of the world.

Going global (again)

'Context' is an important word in ECLAS's central question. Even Christianity, the faith which has been at the forefront of the supposed science/religion debate, does not speak with a single voice. Global context, theological history, and denominational commitments can and do lead to different engagements with science.

A great deal of the scientific theology considered in this book comes from white and English-speaking traditions of the United States and United Kingdom. Reflecting on such contributions, the geographer and historian David Livingstone argues that there is no singular relationship between science and religion, but rather complex discourses in different historical settings and at different locations. As he points out, what this serves is 'to remind us that "science and religion" are always embedded in wider socio-political networks and their relationship is conditioned by the prevailing cultural arrangements.'[24] Theologians Klaas Bom and Benno van den Toren agree, arguing that the debate on science and religion as related to cultural diversity is far more than just a minor issue.[25] Academics are beginning to diversify the science and religion conversation beyond its white, Christian roots.[26]

Even within the United Kingdom, there is a marked difference between Christians in the mainline white majority denominations and the faster-growing Black majority churches. Within the latter, the word 'science' can raise negative images of eugenics, medical experimentation, and technologies within which racism is embedded, stories which earlier chapters of this volume have reasserted.[27] A 2020 poll indicated that twenty-seven per cent of Black Americans reported having a 'great deal' of trust in scientists, compared to forty-four per cent for white adults.[28] Sociologists Cleve Tinsley, Daniel Bolger, and Pamela Prickett, working with Elaine Ecklund, have begun to focus on race and its importance in wider science/faith engagement in the United States, showing how class intersects with race and the perceived utility of science in shaping how Black Protestants in particular regard both science and scientists.[29]

Echoing this practical point, the sociologists Jason Shelton and Michael Emerson have drawn a distinction between 'academic' and 'experiential' approaches to theology.[30] White Protestants, embedded in established institutions, have often taken a more abstract view of theology as a set of correct ideas to be formally debated and passed on to a passive congregation. In this context, pastors and scientists can be viewed as competing authorities. Black Protestants may be more inclined to discuss theology as informal, practical, or needs-based, with an eye toward concrete moral questions. Matters of who is a trustworthy discussion partner can be as important as the rules for discussion. Shelton and Emerson are quick to point out problems which underly their broad, and perhaps simplistic, characterisation. Nevertheless, this insight shines a light on one reason why the 'science–religion' discussion, embedded in academic language and institutions, has sometimes appeared less important to black congregations even while issues of science and religion are front and centre.[31]

This is an area that is evolving quickly, both in terms of the lived experiences of people and the efforts by academics to understand and interpret them. But while several projects have taken this seriously in terms of the sociology and history of science, more attention needs to be paid to the diversity of global theological traditions as they engage with science.[32] In specific fields, such as missiology (the study of Christian mission work), there is growing attention to the importance of distinctive differences in the global context of Christian theology.[33] However, there is as yet little direct reflection on how a global setting for SET might influence the way that Anglo-American culture treats the science/faith interface.

Within the most recent phase of ECLAS's life, it has begun to take these questions seriously, working in collaboration with four international partner hubs based in India, Kenya, Poland, and Singapore.[34] ECLAS is considering the questions: What does it mean to do SET in a global context? And what does such theology mean for the formation and practice of Christian leadership? One part of the answer will, we hope, emerge from a major new survey co-produced by the five hubs (including the United Kingdom).[35] This will involve qualitative and quantitative research on the attitudes of Christian leaders in the five different geographical and cultural settings, their theological and denominational identity, and their national and political influences. It will also, building on our previous work in 'New Narratives', map out an international and cross-cultural array of stories about the mutually constitutive relationships between science and religion.[36] The hubs are developing their own research projects, spinning out into engagement that reinterprets the ECLAS activities outlined above. This global community of enquiry and practice, based on a network of international regional partner hubs, will target the involvement of local leaders, study local needs, and experiment with new models. This is a huge challenge: from variation in resources, language, and history; and through the power structures of church, politics, and education, there are many issues to resolve.

A shared future, based on theologically-informed science and technology, is worth the effort.

Notes

1. IMDb, '2023 Highest Grossing Movies Worldwide', https://www.imdb.com/list/ls562149420/, accessed 17 February 2025.
2. Kai Bird and Martin J. Sherwin, *American Prometheus: The Triumph and Tragedy of J. Robert Oppenheimer* (New York: Knopf, 2005).
3. Richard Polenberg, *In the Matter of J. Robert Oppenheimer: The Security Clearance Hearing* (Ithaca: Cornell University Press, 2002); Jeremy Bernstein, *Oppenheimer: Portrait of an Enigma* (Chicago: Ivan R. Dee, 2004); Charles Thorpe, *Oppenheimer: The Tragic Intellect* (Chicago: University of Chicago Press 2007); Richard Pfau, *No Sacrifice Too Great: The Life of Lewis L. Strauss* (Charlottesville: University Press of Virginia, 1984); Richard Rhodes, *Dark Sun: The Making of the Hydrogen Bomb* (New York: Simon & Schuster, 1985).
4. Ben Sisario, 'How Taylor Swift's Eras Tour Conquered the World', *The New York Times*, 5 August 2023, https://www.nytimes.com/2023/08/05/arts/music/taylor-swift-eras-tour.html, accessed 17 February 2025.
5. Michelle Goldberg, 'The Hunger Fed by "Barbie" and Taylor Swift', *The New York Times*, 24 July 2023, https://www.nytimes.com/2023/07/24/opinion/hunger-barbie-taylor-swift.html, accessed 17 February 2025.
6. Carolyn Y. Johnson, Scott Dance, and Joel Achenbach, 'Here Are the Words Putting Science in the Crosshairs of Trump's Orders', *Washington Post*, 4 February 2025, https://

www.washingtonpost.com/science/2025/02/04/national-science-foundation-trump-executive-orders-words/, accessed 3 March 2025.

7. Colin A. Russell, 'The Conflict of Science and Religion' in *The History of Science and Religion in the Western Tradition: An Encyclopedia*, edited by Gary B. Ferngren, Edward J. Larson, and Darrel W. Amundsen (New York: Routledge, 2000), pp. 12–17. If there was any conflict, it was between the London and Edinburgh medical establishments or between obstetricians and surgeons. The origins of that myth may be located in a footnote in our old friend, Andrew Dickson White, *A History of the Warfare of Science with Theology in Christendom* (New York: Appleton, 1897), p. 263.

8. Arunima Krishna, 'Understanding the Differences between Climate Change Deniers and Believers' Knowledge, Media Use, and Trust in Related Information Sources', *Public Relations Review* 47.1 (2021), unpaginated, https://doi.org/10.1016/j.pubrev.2020.101986; Michael Brüggemann and Sven Engesser, 'Beyond False Balance: How Interpretive Journalism Shapes Media Coverage of Climate Change', *Global Environmental Change* 42 (2017): 58–67, https://doi.org/10.1016/j.gloenvcha.2016.11.004.

9. Climate Action against Disinformation/Institute for Strategic Dialogue, 'Deny, Deceive, Delay: Exposing New Trends in Climate Mis- And Disinformation at COP27 (Vol 2)', n.d., https://caad.info/wp-content/uploads/2023/01/DDD_ExposingClimateDisinfo-COP27.pdf, accessed 17 February 2025.

10. Alexander Michael Petersen, Emmanuel M. Vincent, and Anthony LeRoy Westerling, 'Discrepancy in Scientific Authority and Media Visibility of Climate Change Scientists and Contrarians', *Nature Communications* 10.1 (2019), unpaginated, https://doi.org/10.1038/s41467-019-09959-4.

11. Jeremy Shapiro, *Finding Goldilocks: A Guide for Creating Balance in Personal Change, Relationships, and Politics* (independently published, 2020).

12. Matthew C. Nisbet and Chris Mooney, 'Framing Science', *Science* 316.5821 (2007): 56, DOI: 10.1126/science.1142030.

13. Colin A. Russell, 'The Conflict of Science and Religion', in *Science and Religion: A Historical Introduction*, edited by Gary B. Ferngren (Baltimore: Johns Hopkins University Press, 2002), p. 7. See also Steven Shapin, *The Scientific Revolution* (Chicago: University of Chicago Press, 1996), p. 195; John Hedley Brooke, *Science and Religion: Some Historical Perspectives* (Cambridge: Cambridge University Press, 1991), p. 42.

14. Stillman Drake, *Galileo* (Oxford: Oxford University Press, 1980), p. 64.

15. Charlotte Sleigh, 'Liberation Science-ology: Indigenous and Postcolonial Frameworks for Science and Religion', in *Science and Religion: Approaches from Science and Technology Studies*, edited by Zara Thokozani Kamwendo (London: Palgrave Macmillan, 2024), https://doi.org/10.1007/978-3-031-66387-1_11.

16. https://www.eclasproject.org/, accessed 17 February 2025.

17. Lydia Reid and David Wilkinson, 'Building Enthusiasm and Overcoming Fear: Engaging with Christian Leaders in an Age of Science', *Zygon* 56.4 (2021): 1087–1109, https://doi.org/10.1111/zygo.12731.

18. Kamwendo, Z. T. 2021. "Resistance to Narratives of the Covid-19 Pandemic as an Act of God." *Zygon* 56 (4): 1110–1129.

19. https://www.eclasproject.org/leadership-conferences/, accessed 17 February 2025.

20. https://www.eclasproject.org/climate-cathedral/, accessed 17 February 2025.

21. https://www.eclasproject.org/science-for-seminaries/, accessed 17 February 2025.

22. https://www.eclasproject.org/new-narratives/, access 17 February 2025.

23. Kathryn Pritchard, 'Religion and Science Can Have a True Dialogue', *Nature* 537, no. 7622 (22 September 2016): 451, https://doi.org/10.1038/537451a.

24. John Hedley Brooke and Ronald L. Numbers, *Science and Religion around the World* (New York: Oxford University Press, 2011), p. 287.

25. Klaas Bom and Bernard van den Toren, *Context and Catholicity in the Science and Religion Debate: Intercultural Contributions from French-Speaking Africa* (Leiden: Brill, 2020).

26. For example, Ted Peters, Muzaffar Iqbal, and Syed Nomanul Haq, *God, Life, and the Cosmos: Christian and Islamic Perspectives* (Aldershot: Ashgate, 2002); John Hedley Brooke and Ronald L. Numbers, *Science and Religion around the World* (New York: Oxford University Press, 2011); Michael Fuller, 'Science and Religion in a Global Context', in *Routledge International Handbook of Religion in Global Society*, edited by Jayeel Cornelio, Ateneo de Manila, François Gauthier, Tuomas Martikainen, and Linda Woodhead (Abingdon: Routledge International, 2020), 478–487.

27. Gordon Gauchat, 'The Cultural Authority of Science: Public Trust and Acceptance of Organized Science', *Public Understanding of Science* 20.6 (2011): 751–770, https://doi.org/10.1177/0963662510365246; Terence Keel, *Divine Variations: How Christian Thought Became Racial Science* (Redwood City: Stanford University Press, 2018); Angela Saini, *Superior: The Return of Race Science* (Boston: Beacon Press, 2019).

28. Cary Funk, Brian Kennedy, and Courtney Johnson, 'Trust in Medical Scientists Has Grown in US, but Mainly among Democrats' (Washington, DC: Pew Research Center, 2020), https://www.pewresearch.org/science/2020/05/21/trust-in-medical-scientists-has-grown-in-u-s-but-mainly-among-democrats/, accessed 17 February 2025.

29. Cleve Tinsley, Pamela J. Prickett, and Elaine Howard Ecklund, 'Black Protestant Views of Science', *Du Bois Review: Social Science Research on Race* 15.2 (2018): 533–546, doi:10.1017/S1742058X18000309.

30. Jason E. Shelton and Michael Oluf Emerson, *Blacks and Whites in Christian America* (New York: New York University Press, 2012).

31. We are grateful to Dr Lucas Mix for his input on this section.

32. See, for example, Lisa L. Stenmark, (2021). '"The Benefits of an Entire Civilization": Religion, Science, and Colonialism', in *Religion, Science and Tehcnology in North America: An Introduction*, edited by Lisa L. Stenmark and Whitney A. Bauman (London: Bloomsbury Academic, 2024); Elaine Howard Ecklund, David R. Johnson, Brandon Vaidyanathan, Kirstin R. W. Matthews, Steven W. Lewis, Robert A. Thomson Jr., and Di Di, *Secularity and Science: What Scientists around the World Really Think about Religion* (New York: Oxford University Press, 2019); Renny Thomas, 'Beyond Conflict and Complementarity: Science and Religion in Contemporary India', *Science, Technology and Society* 23.1 (2018): 47–64, https://doi.org/10.1177/0971721817744444.

33. See, for example, Stephen B. Bevans, *An Introduction to Theology in Global Perspective* (New York: Orbis, 2009); Timothy C. Tennent, *Theology in the Context of World Christianity: How the Global Church Is Influencing the Way We Think about and Discuss Theology* (Grand Rapids, Mich.: Zondervan, 2009); Amos Yong, *Renewing Christian Theology: Systematics for a Global Christianity* (Waco, Texas: Baylor University Press, 2014); Jione Havea, *Doing Theology in the New Normal: Global Perspectives* (London: SCM Press, 2021).

34. https://www.eclasproject.org/international-hubs/, accessed 17 February 2025.

35. See, for example, Robert Kozinets, *Netnography: The Essential Guide to Qualitative Social Media Research* (London: Sage, 2019) or Leesa Costello et al., 'Netography: Range of Practices, Misperceptions, and Missed Opportunities', *International Journal of Qualitative Methods*, 16.1 (2017): 1–12, https://doi.org/10.1177/1609406917700647, for a discussion of the importance of including participants as active partners in online qualitative analysis.

36. Michael F. Dahlstrom, 'Using Narratives and Storytelling to Communicate Science with Nonexpert Audiences', *PNAS* 111.4 (2014): 13614–13620, https://doi.org/10.1073/pnas.1320645111; Deserai Crow and Michael Jones 'Narratives as Tools for Influencing Policy Change', *Policy and Politics* 46.2 (2018): 217–234, https://doi.org/10.1332/030557318X15230061022899; Mary E. Hess, 'A New Culture of Learning: Digital Storytelling and Faith Formation', *Dialog* 53.1 (2014): 12–22, https://doi.org/10.1111/dial.12084; Mithra Moezzi, Kathryn B. Janda, and Sea Rotmann, 'Using Stories, Narratives and Storytelling in Energy and Climate Change Research', *Energy Research and Social Science*, 31 (2017): 1–10, https://doi.org/10.1016/j.erss.2017.06.034.

Index

For the benefit of digital users, indexed terms that span two pages (e.g., 52–53) may, on occasion, appear on only one of those pages.